Selected Titles in This Series

743 **Heng Sun,** Spectral decomposition of a covering of $GL(r)$: the Borel case, 2002

742 **J. E. Gilbert, Y. S. Han, J. A. Hogan, J. D. Lakey, D. Weiland, and G. Weiss,** Smooth molecular functions and singular integral operators, 2002

741 **Francisco Santos,** Triangulations of oriented matroids, 2002

740 **Rick Durrett,** Mutual invadability implies coexistence in spatial models, 2002

739 **Georgios K. Alexopoulos,** Sub-Laplacians with drift on Lie groups of polynomial volume growth, 2002

738 **Yasuro Gon,** Generalized Whittaker functions on $SU(2,2)$ with respect to the Siegel parabolic subgroup, 2002

737 **Arjen Doelman, Robert A. Gardner, and Tasso J. Kaper,** A stability index analysis of 1-D patterns of the Gray-Scott model, 2002

736 **Wojciech Chachólski and Jérôme Scherer,** Homotopy theory of diagrams, 2002

735 **Martina Brück, Xi Du, Joonsang Park, and Chuu-Lian Terng,** The submanifold geometries associated to Grassmannian systems, 2002

734 **Michel Van den Bergh,** Blowing up of non-commutative smooth surfaces, 2001

733 **Milé Krajčevski,** Tilings of the plane, hyperbolic groups and small cancellation conditions, 2001

732 **Jan O. Kleppe, Juan C. Migliore, Rosa Miró-Roig, Uwe Nagel, and Chris Peterson,** Gorenstein liaison, complete intersection liaison invariants and unobstructedness, 2001

731 **Jesús Bastero, Mario Milman, and Francisco J. Ruiz,** On the connection between weighted norm inequalities, commutators and real interpolation, 2001

730 **Suhyoung Choi,** The decomposition and classification of radiant affine 3-manifolds, 2001

729 **Michael Grosser, Eva Farkas, Michael Kunzinger, and Roland Steinbauer,** On the foundations of nonlinear generalized functions I and II, 2001

728 **Laura Smithies,** Equivariant analytic localization of group representations, 2001

727 **Anthony D. Blaom,** A geometric setting for Hamiltonian perturbation theory, 2001

726 **Victor L. Shapiro,** Singular quasilinearity and higher eigenvalues, 2001

725 **Jean-Pierre Rosay and Edgar Lee Stout,** Strong boundary values, analytic functionals, and nonlinear Paley-Wiener theory, 2001

724 **Lisa Carbone,** Non-uniform lattices on uniform trees, 2001

723 **Deborah M. King and John B. Strantzen,** Maximum entropy of cycles of even period, 2001

722 **Hernán Cendra, Jerrold E. Marsden, and Tudor S. Ratiu,** Lagrangian reduction by stages, 2001

721 **Ingrid C. Bauer,** Surfaces with $K^2 = 7$ and $p_g = 4$, 2001

720 **Palle E. T. Jorgensen,** Ruelle operators: Functions which are harmonic with respect to a transfer operator, 2001

719 **Steve Hofmann and John L. Lewis,** The Dirichlet problem for parabolic operators with singular drift terms, 2001

718 **Bernhard Lani-Wayda,** Wandering solutions of delay equations with sine-like feedback, 2001

717 **Ron Brown,** Frobenius groups and classical maximal orders, 2001

716 **John H. Palmieri,** Stable homotopy over the Steenrod algebra, 2001

715 **W. N. Everitt and L. Markus,** Multi-interval linear ordinary boundary value problems and complex symplectic algebra, 2001

714 **Earl Berkson, Jean Bourgain, and Aleksander Pełczynski,** Canonical Sobolev projections of weak type $(1,1)$, 2001

(*Continued in the back of this publication*)

Spectral Decomposition of a Covering of $GL(r)$: The Borel Case

MEMOIRS
of the
American Mathematical Society

Number 743

Spectral Decomposition of a Covering of $GL(r)$: The Borel Case

Heng Sun

March 2002 • Volume 156 • Number 743 (fourth of 5 numbers) • ISSN 0065-9266

American Mathematical Society
Providence, Rhode Island

2000 *Mathematics Subject Classification.*
Primary 11F70, 11F72; Secondary 22D12.

Library of Congress Cataloging-in-Publication Data

Sun, Heng, 1968–
Spectral decomposition of a covering of GL(r) : the Borel case / Heng Sun.
p. cm. — (Memoirs of the American Mathematical Society, ISSN 0065-9266 ; no. 743)
"March 2002, volume 156, number 743 (fourth of 5 numbers)."
Includes bibliographical references and index.
ISBN 0-8218-2775-8 (alk. paper)
1. Eisenstein series. 2. Spectral theory (Mathematics) 3. Decomposition (Mathematics) I. Title. II. Series.
QA3.A57 no. 743
[QA295]
510 s—dc21
[515′.243] 2001056089

Memoirs of the American Mathematical Society

This journal is devoted entirely to research in pure and applied mathematics.

Subscription information. The 2002 subscription begins with volume 155 and consists of six mailings, each containing one or more numbers. Subscription prices for 2002 are \$524 list, \$419 institutional member. A late charge of 10% of the subscription price will be imposed on orders received from nonmembers after January 1 of the subscription year. Subscribers outside the United States and India must pay a postage surcharge of \$31; subscribers in India must pay a postage surcharge of \$43. Expedited delivery to destinations in North America \$35; elsewhere \$130. Each number may be ordered separately; *please specify number* when ordering an individual number. For prices and titles of recently released numbers, see the New Publications sections of the *Notices of the American Mathematical Society.*

Back number information. For back issues see the *AMS Catalog of Publications.*

Subscriptions and orders should be addressed to the American Mathematical Society, P. O. Box 845904, Boston, MA 02284-5904. *All orders must be accompanied by payment.* Other correspondence should be addressed to Box 6248, Providence, RI 02940-6248.

Memoirs of the American Mathematical Society is published bimonthly (each volume consisting usually of more than one number) by the American Mathematical Society at 201 Charles Street, Providence, RI 02904-2294. Periodicals postage paid at Providence, RI. Postmaster: Send address changes to Memoirs, American Mathematical Society, P. O. Box 6248, Providence, RI 02940-6248.

This publication is indexed in *Science Citation Index*®, *SciSearch*®, *Research Alert*®, *CompuMath Citation Index*®, *Current Contents*®/*Physical, Chemical & Earth Sciences.*
Printed in the United States of America.

∞ The paper used in this book is acid-free and falls within the guidelines established to ensure permanence and durability.
Visit the AMS home page at URL: `http://www.ams.org/`

10 9 8 7 6 5 4 3 2 1 07 06 05 04 03 02

Contents

Abstract

Let F be a number field and $\mathbf{A}$ the ring of adeles over F. Suppose $\overline{G(\mathbf{A})}$ is a metaplectic cover of $G(\mathbf{A}) = GL(r, \mathbf{A})$ which is given by the n-th Hilbert symbol on $\mathbf{A}$. According to Langlands' theory of Eisenstein series, the decomposition of the right regular representation on $L^2\left(G(F)\backslash\overline{G(\mathbf{A})}\right)$ can be understood in terms of the residual spectrum of Eisenstein series associated with cuspidal data on standard Levi subgroups $\overline{M}$. Under an assumption on the base field F, this paper calculates the spectrum associated with the diagonal subgroup $\overline{T}$. Specifically, the diagonal residual spectrum is at the point $\lambda = ((r-1)/2n, (r-3)/2n, \cdots, (1-r)/2n)$. Each irreducible summand of the corresponding representation is the Langlands quotient of the space induced from an irreducible automorphic representation of $\overline{T}$, which is invariant under symmetric group $\mathfrak{S}_r$, twisted by an unramified character of $\overline{T}$ whose exponent is given by λ.

Received by the editor August 17, 1998.

1991 *Mathematics Subject Classification.* Primary 11F70, 11F72; Secondary 22D12.

Key words and phrases. spectral decomposition, representation, metaplectic group, metaplectic form, intertwining operator.

Introduction

In sixties, Langlands presented a general theory on how to decompose the regular representations of a large category of groups including reductive groups, modulo the understanding of the space of cuspidal forms ([**Lan76**] and [**MW93**]). Using this theory, Mœglin and Waldspurger gave an explicit decomposition for the general linear group [**MW89**]. On the other hand, Kazhdan and Patterson initiated the study of automorphic representations of metaplectic covers of the general linear groups [**KP84**]. It is then natural to ask for the spectral decomposition for metaplectic groups. This paper considers the spectrum associated with the Borel subgroup.

Let F be a number field and $\mathbf{A}$ the ring of adeles over F. Suppose $\overline{G(\mathbf{A})}$ is a metaplectic cover of $G(\mathbf{A}) = GL(r, \mathbf{A})$. It is a central extension of $G(\mathbf{A})$ by a finite abelian group μ which splits over $G(F)$. Let $p : \overline{G(\mathbf{A})} \to G(\mathbf{A})$ be the natural projection. Denote by H the subgroup of $GL(r)$ consisting of diagonal matrices with determinant 1. Then $\mu_n = [\overline{H(\mathbf{A})}, \overline{H(\mathbf{A})}]$ is a cyclic group of order n. This order n is more important than the covering number of $\overline{G(\mathbf{A})}$. The base field F necessarily contains the group of all n-th roots of unity which we identify with μ_n.

Let $\mathfrak{Z}(\overline{G}(\mathbf{A}))$ be the center of $\overline{G}(\mathbf{A})$ and χ be a character of $\mathfrak{Z}(\overline{G}(\mathbf{A}))$ whose restriction to μ_n is a fixed embedding $\mu_n \hookrightarrow \mathbf{C}^\times$. Consider the space

$$L^2(\overline{G}, \chi) = \left\{ f : f(zg) = \chi(z)f(g), \forall z \in \mathfrak{Z}(\overline{G}(\mathbf{A})), |f| \in L^2\left(\mathfrak{Z}(\overline{G}(\mathbf{A}))G(F)\backslash\overline{G(\mathbf{A})}\right)\right\}.$$

Suppose $M = GL(r_1) \times \cdots \times GL(r_k)$ is a standard Levi subgroup and ρ is an irreducible cuspidal automorphic representation of $\overline{M(\mathbf{A})}$ whose restriction to $\mathfrak{Z}(\overline{G}(\mathbf{A}))$ is χ. Let V_ρ be the isotropic subspace of ρ in $L^2(\overline{M}, \chi_\rho)$ where χ_ρ is the central character of ρ. If

$$\underline{s} \in X(M) = \left\{ \underline{s} \in \mathbf{C}^k : \sum_{i \le k} r_i s_i = 0 \right\},$$

denote by $\rho[\underline{s}]$ the representation of $\overline{M(\mathbf{A})}$ defined by

$$\rho[\underline{s}](m) = \rho(m) \prod_i |\det(g_i)|^{s_i}.$$

Define $\varrho = \{\rho[\underline{s}], \underline{s} \in X(M)\}$. We call a pair (M, ϱ) a cuspidal datum. Two cuspidal data are equivalent if they are conjugate by an element $\sigma \in GL(F)$. According to Langlands' theory of Eisenstein series, associated with each equivalence class of cuspidal data Θ, there is a subspace $L^2\left(\overline{G}, \chi\right)_\Theta$. And

$$L^2\left(\overline{G}, \chi\right) = \oplus L^2\left(\overline{G}, \chi\right)_\Theta$$

where Θ runs through all equivalence classes of cuspidal data (M, ϱ). In this paper, we consider the diagonal case, i.e., when M is the diagonal subgroup T and $\Theta_0 = (T, \varrho)$ where ϱ is an equivalence class (under the conjugation by the Weyl group and the translation by $\underline{s} \in X(T)$) of irreducible automorphic representations of $\overline{T}$. Fix such a ϱ.

For each place v on F, let $(\cdot,\cdot)_v$ be the n-th Hilbert symbol on F_v. We need the following assumption on the base field F:

ASSUMPTION 0.1. For each place v, $(-1,-1)_v = 1$.

Observe this assumption is satisfied when F contains the group of $2n$-th roots of unity or -1 is an n-th power. In particular, this is the case when n is odd.

Let $\mathfrak{S}_r$ be the symmetric group on r letters. Identify $\mathfrak{S}_r$ with the quotient group of the normalizer of $T(F)$ in $G(F)$ by $T(F)$. A partition $\underline{p}$ of r is finite sequence $\underline{p} = (p_1, \cdots, p_m)$ satisfying $p_1 + \cdots + p_m = r$. Associated with each partition, there is a set of intervals $\Delta_i = \{j \in \mathbf{Z} : p_1 + \cdots + p_{i-1} + 1 \le j \le p_1 + \cdots + p_i\}$, $1 \le i \le m$. Let $\mathfrak{S}(\underline{p})$ be the subset of $\mathfrak{S}_r$ consisting of permutations stabilizing the intervals, i.e., $\mathfrak{S}(\underline{p}) = \{\sigma \in \mathfrak{S}_r : \sigma(j) \in \Delta_i, \forall i \le m, j \in \Delta_i\}$. Denote by $\mathbf{P}(\varrho)$ the set of all the pairs $(\underline{p}, \pi)$, where $\underline{p}$ is a partition of r and $\pi = \pi_{\underline{p}} \in \varrho$ such that $\sigma(\pi_{\underline{p}}) = \pi_{\underline{p}}$ for all $\sigma \in \mathfrak{S}(\underline{p})$. For each $(\underline{p}, \pi) \in \mathbf{P}(\varrho)$ define a set with multiplicities $\{(\underline{p}, \pi)\} = \{(\tau\underline{p}, \tau(\pi)) : \tau \in \mathfrak{S}_m\}$. Here we identify an element in $\mathfrak{S}_m$ as an element in $\mathfrak{S}_r$ permuting the intervals Δ_k and increasing on each of them. Denote by $\mathbf{P}^0(\varrho)$ the collection of sets with multiplicities thus obtained.

Let M is the standard Levi subgroup associated with $\underline{p}$, i.e., $M = GL(p_1) \times \cdots \times GL(p_m)$. Denote by $J_{(\underline{p},\pi)}$ the unique irreducible quotient of $i_{\overline{T}}^{\overline{M}} \pi_{\underline{p}}[\lambda(\underline{p})]$ where

$$\lambda(\underline{p}) = \\ ((p_1 - 1)/2n, (p_1 - 3)/2n, \cdots, (1 - p_1)/2n, (p_2 - 1)/2n, \cdots, (1 - p_r)/2n).$$

Let $Z(\underline{p}) - \{\underline{s} \in i\mathbf{R}^m : \sum p_i s_i = 0\}$. Denote by $\mathfrak{S}(\underline{p}, \pi)$ the set of all $\tau \subset \mathfrak{S}_m$ such that $\tau\pi = \pi$. The group $\mathfrak{S}(\underline{p}, \pi)$ acts on $Z(\underline{p})$ by permuting coordinates. The main Theorem 3.15 asserts that

$$\begin{aligned} L^2(\overline{G}, \chi)_{\Theta_0} &= \oplus_{\{(\underline{p},\pi)\} \in \mathbf{P}^0(\varrho)} L_{\{(\underline{p},\pi)\}}, \\ L_{\{(\underline{p},\pi)\}} &= m(\overline{T}) \int_{Z(\underline{p})/\mathfrak{S}(\underline{p},\pi)}^{\oplus} i_{\overline{M}}^{\overline{G}} \left(J_{(\underline{p},\pi)}[\underline{t}]\right) d\underline{t}, \end{aligned}$$

where $m(\overline{T})$ is the multiplicity. In particular, suppose for some unitary $\rho \in \varrho$, we have $\sigma\rho = \rho$, for all $\sigma \in \mathfrak{S}_r$. Then the discrete part of $L^2(\overline{G}, \chi)_{\Theta_0}$ is certain multiple of the unique irreducible quotient of $i_{\overline{T}}^{\overline{G}} \rho[\underline{\lambda}]$ where

$$\underline{\lambda} = ((N-1)/2n, (N-3)/2n, \cdots, (1-N)/2n).$$

Otherwise, the discrete part is 0.

Remark that if the covering group is defined as in [**KP84**, §0.2], the above multiplicity $m(\overline{T})$ is 1.

To establish the above result, first notice that the general theory on Eisenstein theory applies to metaplectic groups. This is explicitly stated in [**MW93**]. Also, since a local metaplectic group is locally compact, all theories about general locally compact groups apply. In particular, Frobenius reciprocity and the theorem of Bernstein-Zelevinski on Jacquet modules [**BZ77**, 5.2 Theorem] hold. Note due to Assumption 0.1, we do not need to study real groups when we study local representations.

We follow [**MW89**] to calculate the residues of Eisenstein Series and the associated representations. The main obstruction is that a Levi subgroup $\overline{M(\mathbf{A})}$ of $\overline{G(\mathbf{A})}$ is not a product of coverings of smaller reductive groups. So we are going to consider a bigger category of groups $\overline{M}$ and check the method of Mœglin and Waldspurger works through for $\overline{M}$.

The first chapter is to normalize the local intertwining operator and derive some consequences. The key Lemma 1.15 is [**KP84**, Theorem 1.2.6] in a general situation. It is proved by direct calculations without using the global method.

The second chapter studies the local intertwining operators. It corresponds to [**MW89**, Part I] and the basic idea is similar. However there seems no easy way to build irreducible representations of M from those of $\overline{GL(k)}$, like what tensor products we use in linear group case. Also the irreducible dual of $\overline{GL(k)}$ is not completely understood.

The third chapter derives the main result as stated in this introduction. The representations of the diagonal subgroup is also studies. We don't have multiplicity one for a general metaplectic group.

The last chapter proves the principal lemma left from Chapter Three. The method is of Mœglin and Waldspurger [**MW89**].

The following is a list of commonly used notations:

$$\mathbf{N},\ \ \mathbf{Z},\ \ \mathbf{Q},\ \ \mathbf{R},\ \ \mathbf{C},\qquad \text{and}\qquad \frac{1}{n}\mathbf{Z} = \{k \in \mathbf{R} :\ nk \in \mathbf{Z}\}\,.$$

Temporary notation used only within a section are not listed.

Acknowledgments

This paper is based on my Ph.D. thesis under the supervision of James Arthur, whose insight and guidance are greatly appreciated. I am indebted to Jeff Adams, Brooks Roberts and Paul Mezo for many useful discussions. I would also like to thank the thesis examiners Joe Repka and Freydoon Shahidi for reading the thesis and making corrections.

CHAPTER 1

Preliminaries

1.1. The Group

Let F be a local field and $G = GL(r,F)$. Let $\overline{G}$ be a covering group of G by a finite abelian group μ. This means a totally disconnected group (resp. Lie group) $\overline{G}$ when F is a p-adic field (resp. archimedean field), together with the following short exact sequence

$$1 \longrightarrow \mu \longrightarrow \overline{G} \xrightarrow{p} G \longrightarrow 1,$$

such that $p : \overline{G} \to G$ is a topological quotient map. For any subset L of G, denote $\overline{L} = p^{-1}(L)$.

Let N be the upper-triangular unipotent subgroup of G. Let $1 \leq i \neq j \leq r$. Identify the pair (ij) with the simple transposition interchanging i and j. Denote by $n'_{ij}(x)$ the matrix obtained from the identity matrix with the (i,j) entry replaced by x. Then N is generated by $\{n'_{ij}(x) : i < j, x \in F\}$. By [**Ste62**], the covering $p : \overline{G} \to G$ splits over N, i.e., there is a homomorphism $s : N \to \overline{N}$ such that $ps = 1_N$. Identify N with a subgroup of $\overline{G}$ via s. Denote

$$\begin{aligned}
n_{ij}(x) &= s(n'_{ij}(x));\\
w'_{ij}(x) &= n'_{ij}(x)\, n'_{ji}(-x^{-1})\, n'_{ij}(x);\\
w_{ij}(x) &= n_{ij}(x)\, n_{ji}(-x^{-1})\, n_{ij}(x)\\
h'_{ij}(x) &= w'_{ij}(x)\, w'_{ij}(-1);\\
h_{ij}(x) &= w_{ij}(x)\, w_{ij}(-1).
\end{aligned}$$

Let $\mathfrak{S}_r$ be the group of all permutations on r letters. Any element σ can be written as $\sigma = \sigma_1 \cdots \sigma_k$ where σ_i are simple transpositions and k is the length of σ. Denote

$$w'_\sigma = w'_{\sigma_1}(1) \cdots w'_{\sigma_k}(1); \tag{1.1}$$

$$w_\sigma = w_{\sigma_1}(1) \cdots w_{\sigma_k}(1). \tag{1.2}$$

Remark that both w'_σ and w_σ are independent of the decomposition of $\sigma = \sigma_1 \cdots \sigma_k$ [**Mat69**, p44]. Identify σ with w_σ. Let $W' = \{w'_\sigma : \sigma \in \mathfrak{S}_r\}$ and $W = \{w_\sigma : \sigma \in \mathfrak{S}_r\}$. Define a map $s : W' \to W$ by

$$s(w'_\sigma) = w_\sigma.$$

Observe that $h'_{ij}(x)$ is the diagonal matrix with x and x^{-1} at the i-th and j-th entries respectively and 1 at the other diagonal entries. Also, the group H of diagonal matrices with determinant 1 is generated by the set $\{h'_{ij}(x) : i < j, x \in F\}$. Define a section $s : H \to \overline{H}$ for the covering $p : \overline{H} \to H$ by

$$s(h'_{12}(x_1)h'_{23}(x_2) \cdots h'_{r-1,r}(x_{r-1})) = h_{12}(x_1)h_{23}(x_2) \cdots h_{r-1,r}(x_{r-1}).$$

Let T be the group of diagonal matrices and T_1 be the group of diagonal matrices with 1 at each diagonal entry except the first entry. The $T = HT_1$. Let $s : T_1 \to \overline{T_1}$ be a section such that the associated 2-cocycle

$$\sigma : T_1 \times T_1 \to \mu, \qquad \sigma(t,t') = s(t)s(t')s(tt')^{-1}, \qquad \forall t,t' \in T_1$$

is a constant in a neighborhood of identity in the group $T_1 \times T_1$. The existence of such a section is given by [**Nag49**, Theorem 2].

By the decomposition $G = SL(r)T_1$ and the Bruhat decomposition for $SL(r,F)$, we can define a section $s : G \to \overline{G}$ as follows:

$$s(n_1hwn_2t_1) = s(n_1)s(h)s(w)s(n_2)s(t_1), \qquad \forall n_1,n_2 \in N, h \in H, w \in W, t_1 \in T_1.$$

The above definition does not depend on the particular decomposition of $n_1hwn_2t_1$.

Let σ be the 2-cocycle associated with the above defined section, i.e.,

$$\sigma(g,g') = s(g)s(g')s(gg')^{-1}, \qquad \forall g,g' \in G.$$

Then the function

$$c : F^\times \times F^\times \to \mu, \qquad c(x,y) = \sigma(h_{ij}(x), h_{ij}(y)), \qquad \forall x,y \in F^\times$$

is a bilinear topological Steinberg cocycle [**Moo68**]. It is a continuous function $c : F^\times \times F^\times \to \mu$ such that for any $x,y,z \in F^\times$,

$$\begin{aligned} c(x,y) &= c(y,x)^{-1}, \\ c(x,y)c(z,y) &= c(xz,y), \\ c(x,1-x) &= 1, \\ c(x,-x) &= 1. \end{aligned}$$

There is an integer n such that the image of c is a finite cyclic group, denoted μ_n, of order n. Furthermore, F contains the group of all n-th roots of unity which we identify with μ_n. Fix such an integer n all through this paper unless otherwise stated. The cocycle σ can be calculate explicitly on H:

$$\sigma(h,h') = \prod_{i \le j} c(h_i, h'_j), \qquad \forall h = \mathrm{diag}(h_i), h' = \mathrm{diag}(h'_i) \in H.$$

The following two identities are needed in future calculations:

$$w_{ij}(-1)n_{ij}(x) = n_{ij}(-x^{-1})h_{ij}(x^{-1})n_{ji}(x^{-1}), \tag{1.3}$$

and

$$\begin{aligned} &w_{ij}(-1)n_{ij}(y)w_{ij}(1)n_{ij}(x) = \\ &n_{ij}(x(1-xy)^{-1})h_{ij}((1-xy)^{-1})n_{ji}(y(1-xy)^{-1})c(1-xy,-x). \end{aligned} \tag{1.4}$$

They can be easily deduced from [**Mat69**, Lemme 5.2].

If L is a group, denote by $\mathfrak{Z}(L)$ the center of the group L. For example, $\overline{\mathfrak{Z}(G)}$ is the group of all element in $\overline{G}$ whose projections are scalar matrices. Remark that $\overline{\mathfrak{Z}(G)} \neq \mathfrak{Z}(\overline{G})$. Define a map

$$\iota : F^\times \to \overline{T_1}, \quad \iota(x) = s(\mathrm{diag}(x,1,\cdots,1)). \tag{1.5}$$

According to [**Sun**, (17)],

$$\iota(x)w_{12}(1) = h_{12}(x)w_{12}(1)\iota(x). \tag{1.6}$$

If g and h are two elements in a group, let $[g,h]=ghg^{-1}h^{-1}$. For any two subgroups L_1 and L_2 of G such that $[L_1,L_2]=1$, observe that $[s(l_1),s(l_2)]$ is a bicharacter on $L_1\times L_2$.

LEMMA 1.1. *For any* $\alpha\in\overline{\mathfrak{Z}(G)}$, $n\in\overline{N}$ *and* $w_\tau,\tau\in\mathfrak{S}_r$, *we have* $[\alpha,n]=[\alpha,w_\tau]=1$. *So*

$$\mathfrak{Z}(\overline{G})=\overline{\mathfrak{Z}(G)}\cap\mathfrak{Z}(\overline{T}).$$

PROOF. By the remark before the lemma, it is enough to show that

$$[\alpha,w_{i,i+1}(1)]=1 \tag{1.7}$$

for any $i<r$. This follows from the following two identities: for any $x\in F^\times$,

$$[s(\mathrm{diag}(1,\cdots,x,\cdots,1)),w_{i,i+1}(1)] = 1; \tag{1.8}$$

$$[s(\mathrm{diag}(1,\cdots,1,x,x,1,\cdots,1)),w_{i,i+1}(1)] = 1; \tag{1.9}$$

where the x in (1.8) is at the j-th entry with $j\neq i,i+1$, and the the two x's in (1.9) are at the i-th and $i+1$-th positions. Now (1.8) follows from

$$[\iota(x),w_{i,i+1}(1)] = 1; \tag{1.10}$$

$$[h_{kl}(x),w_{i,i+1}(1)] = 1, \tag{1.11}$$

where $\{k,l\}\cap\{i,i+1\}=\emptyset$. The above two identities follows from [**Sun**, (16)] and [**Mat69**, Lemme 5.2, (d)] respectively.

For (1.9), if $i\neq 1$, write

$$s(\mathrm{diag}(1,\cdots,1,x,x,1,\cdots,1))=h_{1,i}(x^{-2})h_{i,i+1}(x^{-1})\iota(x^2).$$

Then (1.9) follows from (1.10) and [**Mat69**, Lemme 5.2, (d)]. The proof of (1.9) is easier for the case $i=1$. □

Now suppose F is a p-adic field. Let $\mathfrak{O}$ be the ring of integers and ϖ be a fixed generator of the prime idea of $\mathfrak{O}$.

The section s has the following properties. There is an open subgroup U of G such that the restriction of s on U is an isomorphism, both topologically and algebraically. There is an integer l such that for any (i,j), the subgroup $\{n_{ij}(x): x\in\varpi^l\mathfrak{O}\}$ is contained U. Let $K=\{k\in GL(r,\mathfrak{O}):\det(k)\in\mathfrak{O}^\times\}$, which is a maximal compact subgroup of G. The $s(U)$ is a normal subgroup of $\overline{K}$ [**Sun**].

When $|n|_F=1$ and the residue field of F contains at least 4 elements, then $\overline{G}$ splits over K, i.e., there is an isomorphism $\kappa:K\to\overline{G}$ such that $p\kappa=1_K$. Let $K^*=\kappa(K)$. By[**Sun**, Lemma 6], for $i<j$, we have

$$w_{ij}(1),w_{ij}(-1)\in K^*. \tag{1.12}$$

We close this section by a remark on the covering groups over an Archimedean field F. By [**Moo68**, Lemma (3.4)], $H^2(F^\times,\mu)$ is trivial. So [**Sun**, Proposition 1] implies that $\overline{G}=\overline{GL(r,F)}$ is uniquely determined by its restriction to $\overline{G'}=\overline{SL(r,F)}$. In conclusion, modulo trivial coverings of groups, there are only three groups over archimedean fields to be considered: $GL(r,\mathbf{C})$, $GL(r,\mathbf{R})$ and $\overline{GL(r,\mathbf{R})}$ which is the nontrivial 2-fold covering of $GL(r,\mathbf{R})$ given by the nontrivial Steinberg cocycle on $\mathbf{R}$.

1.2. The Diagonal Subgroup and Its Representations

As before, denote by T the subgroup of G consisting of diagonal matrices. Observe that $[s(\cdot), s(\cdot)]$ is a bilinear function on $T \times T$. The following lemma follows from this observation and a simple calculation.

LEMMA 1.2. *If $a = (a_i), b = (b_i) \in T$, then*

$$[s(a), s(b)] = [\iota(\det(a)), \iota(\det(b))]c(\det(a), \det(b))^{-1} \prod_i c(a_i, b_i). \tag{1.13}$$

Let $\overline{C}$ be the center of $\overline{T}$. Then $\overline{T}/\overline{C}$ is a finite abelian group by the above lemma.

LEMMA 1.3. *An irreducible representation of $\overline{T}$ is uniquely determined by its central character. Specifically, let $\overline{A}$ be a maximal abelian subgroup of $\overline{T}$.*

1) The restriction of any irreducible representation of $\overline{T}$ to $\overline{C}$ is a direct sum of $|T/A|$ copies of a character of $\overline{C}$.

2) If χ is a character of $\overline{C}$, then the representation of $\overline{T}$ induced from a character of $\overline{A}$ which extends χ is irreducible. Furthermore, this representation does not depend on the choice of the maximal abelian subgroup $\overline{A}$ and how χ extends to $\overline{A}$.
□

If L is an abelian group, denote $L^n = \{l^n : l \in L\}$. For example, $F^{\times n}$, $\mathfrak{O}^n$ and H^n are going to be used frequently. Let B be a maximal subgroup of $F^\times$ subject to the following property:

$$(b, b') = 1, \qquad \forall b, b' \in B. \tag{1.14}$$

Let

$$\hat{B} = \{x \in F^\times : c(x, b) = 1, \forall b \in B\}.$$

LEMMA 1.4. *The two groups B and $\hat{B}$ are not equal to each other if and only if $c(-1, -1) = -1$. If this is the case, $-1 \in \hat{B} - B$ and $\hat{B}/B \cong \{\pm 1\}$.*

PROOF. First observe that if $x \in \hat{B}$ and $(x, x) = 1$, then $x \in B$ by the maximality of B. Also we have $c(b, b) = 1, \forall b \in B \Rightarrow c(b, -1) = 1, \forall b \in B \Rightarrow -1 \in \hat{B}$.

Now the first statement follows from the following equivalences: $c(-1, -1) = 1 \Leftrightarrow -1 \in B \Leftrightarrow (x, -1) = 1, \forall x \in \hat{B} \Leftrightarrow (x, x) = 1, \forall x \in \hat{B} \Leftrightarrow \hat{B} = B$.

If $c(-1, -1) = -1$, then for any $x \in \hat{B} - B$, $(-x, -x) = -(x, x) = 1$. So $-x \in B$. □

LEMMA 1.5.

$$|B/F^{\times n}| = |F^\times/\hat{B}| = \begin{cases} \frac{n}{|n|_F^{1/2}} & \text{if } (-1,-1) = 1 \\ \frac{n}{(2|n|_F)^{1/2}} & \text{if } (-1,-1) = -1 \end{cases}.$$

PROOF. The following two injections

$$F^\times/\hat{B} \to (B/F^{\times n})^* \qquad x \mapsto (x, \cdot);$$
$$B/F^{\times n} \to \left(F^\times/\hat{B}\right)^* \qquad b \mapsto (b, \cdot).$$

imply that

$$|F^\times/\hat{B}| \le |\left(B/F^{\times n}\right)^*| = |B/F^{\times n}| \le |\left(F^\times/\hat{B}\right)^*| = |F^\times/\hat{B}|.$$

Now the lemma follows from the formula $|F^\times/F^{\times n}| = n^2/|n|_F$ and Lemma 1.4. □

Let

$$H_{ij}(B) = \{h_{ij}(b) : b \in B\}, H_{ij}(\hat{B}) = \{h_{ij}(b) : b \in \hat{B}\}.$$

It follows from (1.13) that $\overline{H_{ij}(\hat{B})}$ is an abelian group. Furthermore, the following relations hold. Let m be the order of μ.

$$\begin{aligned} \overline{C} &\supset \overline{T^m}, && (1.15)\\ \overline{C} &\subset \mathfrak{Z}(\overline{G})\overline{T^n}, && (1.16)\\ \overline{C} \cap \overline{H} &= \overline{H^n}, && (1.17) \end{aligned}$$

To prove (1.16), let $c \in \overline{C}$. Then it commutes with $s(h_{ij}(x))$ for any $x \in F^\times$. Apply (1.13) to c and $s(h_{ij}(x))$, we get (1.16). Identity (1.17) follows similarly.

It follows from (1.17) that

LEMMA 1.6. *The center $\overline{C}$ of $\overline{T}$ is an open subgroup $\overline{T}$.* □

In the following calculations, we are going to use [**Mat69**, Lemme 5.2] and (1.6) repeatedly.

LEMMA 1.7. *Let $a \in \overline{T}$ such that the i-th and $(i+1)$-th entries of $p(a)$ are 1. Then $[a, w_{i,i+1}] = 1$.*

PROOF. Observe $[\cdot, w_{i,i+1}]$ is a character on the group $\{a \in \overline{T} : p(a)_i = p(a)_{i+1} = 1\}$. So modulo some trivial identities, we need to show for

$$a = s(\mathrm{diag}(1, \cdots, 1, x, 1, \cdots, 1))$$

with the x at the $(i+2)$-th entry,

$$[w_{i,i+1}, a] = 1. \qquad (1.18)$$

If $i = 1$, then $a = h_{12}(x^{-1})h_{23}(x^{-1})\iota(x)$ and

$$\begin{aligned} w_{12}a &= w_{12}h_{12}(x^{-1})h_{23}(x^{-1})\iota(x)\\ &= h_{12}(x)w_{12}h_{23}(x^{-1})\iota(x)\\ &= h_{12}(x)h_{23}(x^{-1})h_{12}(x^{-1})w_{12}\iota(x)\\ &= h_{12}(x)h_{23}(x^{-1})h_{12}(x^{-1})^2\iota(x)w_{12}c(x,-1)\\ &= h_{23}(x^{-1})h_{12}(x^{-1})\iota(x)w_{12}c(x,-1)\\ &= aw_{12}. \end{aligned}$$

The proof of (1.18) is easier if $i \neq 1$. □

LEMMA 1.8. *Let $a \in \overline{T}$ and $1 \le i < j \le r$. Suppose $p(a) = \mathrm{diag}(a_1, \cdots, a_r)$. Then*

$$a^{-1}w_{ij}^{-1}aw_{ij} = h_{ij}(a_i^{-1}a_j)c(a_ia_j, -1).$$

PROOF. We first prove the lemma when $j = i + 1$. By Lemma 1.7, we only need to show that $w_{i,i+1}$ commutes with $a = s(\mathrm{diag}(1, \cdots, 1, x, y, 1, \cdots, 1))$, where the x is at the i-th entry.

First consider that case when $i = 1$.

$$\begin{aligned}
&a^{-1}w_{12}^{-1}aw_{12}\\
&= \left(h_{12}(y^{-1})\iota(xy)\right)^{-1} w_{12}^{-1}h_{12}(y^{-1})\iota(xy)w_{12}\\
&= \left(\iota(xy)h_{12}(y^{-1})c(y^{-1},xy)\right)^{-1} w_{12}^{-1}h_{12}(y^{-1})w_{12}h_{12}(x^{-1}y^{-1})\iota(xy)\\
&= h_{12}(y^{-1})^{-1}\iota(xy)^{-1}h_{12}(y)h_{12}(x^{-1}y^{-1})\iota(xy)c(y,xy)\\
&= h_{12}(y^{-1})^{-1}\iota(xy)^{-1}h_{12}(x^{-1})\iota(xy)\\
&= h_{12}(y^{-1})^{-1}h_{12}(x^{-1})c(x^{-1},xy)\\
&= h_{12}(yx^{-1})c(y^{-1},yx^{-1})^{-1}c(x^{-1},xy)\\
&= h_{12}(yx^{-1})c(-1,xy).
\end{aligned}$$

If $i \neq 1$, then $a = h_{1i}(x^{-1}y^{-1})h_{i,i+1}(y^{-1})\iota(xy)$.

$$\begin{aligned}
&a^{-1}w_{i,i+1}^{-1}aw_{i,i+1}\\
&= \left(\iota(xy)h_{1i}(x^{-1}y^{-1})h_{i,i+1}(y^{-1})c(xy,-1)\right)^{-1}\\
&\quad \cdot w_{i,i+1}^{-1}\left(h_{1i}(x^{-1}y^{-1})h_{i,i+1}(y^{-1})\iota(xy)\right) w_{i,i+1}\\
&= \left(h_{1i}(x^{-1}y^{-1})h_{i,i+1}(y^{-1})\right)^{-1} \iota(xy)^{-1}h_{1,i+1}(x^{-1}y^{-1})h_{i,i+1}(y)\iota(xy)\\
&= \left(h_{1i}(x^{-1}y^{-1})h_{i,i+1}(y^{-1})\right)^{-1} h_{1,i+1}(x^{-1}y^{-1})h_{i,i+1}(y)c(xy,-1)\\
&= h_{i,i+1}(x^{-1}y)c(xy,-1).
\end{aligned}$$

So the lemma is proved for $j = i+1$.

Now if $j > i+1$, then

$$w_{ij} = w_{j-1,j}w_{i,j-1}w_{j-1,j} = w_{j-1,j}w_{i,j-1}w_{j-1,j}^{-1}h_{j-1,j}(-1)^{-1}.$$

Assuming the lemma is true for $j-1$, we calculate for j:

$$\begin{aligned}
&a^{-1}w_{ij}^{-1}aw_{ij}\\
&= a^{-1}h_{j-1,j}(-1)\left(w_{j-1,j}w_{i,j-1}w_{j-1,j}^{-1}\right)^{-1} a\left(w_{j-1,j}w_{i,j-1}w_{j-1,j}^{-1}\right) h_{j-1,j}(-1)^{-1}\\
&= h_{j-1,j}(-1)\left(\left[a^{-1},\left(w_{j-1,j}w_{i,j-1}w_{j-1,j}^{-1}\right)^{-1}\right]\right) h_{j-1,j}(-1)^{-1}c(-1,a_{j-1}a_j)\\
&= h_{j-1,j}(-1)w_{j-1,j}\left[w_{j-1,j}^{-1}a^{-1}w_{j-1,j},w_{i,j-1}^{-1}\right] w_{j-1,j}^{-1}h_{j-1,j}(-1)^{-1}c(a_ia_j,-1)\\
&= h_{j-1,j}(-1)w_{j-1,j}h_{i,j-1}(a_i^{-1}a_j)w_{j-1,j}^{-1}h_{j-1,j}(-1)^{-1}\\
&= h_{j-1,j}(-1)h_{i,j-1}(a_i^{-1}a_j)h_{j-1,j}(a_i^{-1}a_j)h_{j-1,j}(-1)^{-1}\\
&= h_{ij}(a_i^{-1}a_j)c(a_ia_j,-1).
\end{aligned}$$

The lemma then follows by induction. □

If ρ is an irreducible representation of $\overline{T}$, let χ_ρ be its central character of ρ. For $i,j \leq r$, define a quasi-character ρ_{ij}^n of $F^\times$ by

$$\rho_{ij}^n(x) = \chi_\rho(h_{ij}(x^n)), \quad \forall x \in F^\times.$$

If $i < j \leq r$, by convention, we identify (ij) with w_{ij}. So $(ij)\rho = w_{ij}\rho$ is the representation of $\overline{T}$ given by $w_{ij}\rho(t) = \rho(w_{ij}^{-1}tw_{ij})$, $\forall t \in \overline{T}$. The following discussions concern when $G = GL(r)$ for $r \geq 3$.

LEMMA 1.9. *Assume $r \geq 3$ and $i,j \leq r$. Let ρ be an irreducible representation of $\overline{T}$. Then $w_{ij}(\rho) = \rho$ if and only if $\rho_{ij}^n = 1$.*

PROOF. Suppose χ_ρ is the central character of ρ. By Lemma 1.3, we need to show $w_{ij}(\chi_\rho) = \chi_\rho$ if and only if $\rho_{ij}^n = 1$. We first show that

$$s(H_{ij}(F^{\times n})) = \{w_{ij}^{-1}c^{-1}w_{ij}c : c \in \overline{C}\}. \tag{1.19}$$

Observe that if $c \in \overline{C}$, then by (1.16), $c = zt^n\xi$ for some $z \in \overline{\mathfrak{Z}(G)}$, $t \in \overline{T}$ and $\xi\mu$. So

$$w_{ij}^{-1}c^{-1}w_{ij}c = w_{ij}^{-1}(zt^n)^{-1}w_{ij}zt^n = w_{ij}^{-1}t^{-n}w_{ij}t^n \in s(H^n).$$

We used Lemma 1.1 in the above calculation.

Conversely, let $h_{ij}(x^n) \in s(H_{ij}(F^{\times n}))$ and $k \leq r, k \neq i, j$. Then

$$h_{ij}(x^n) = w_{ij}^{-1}h_{ik}(x^n)^{-1}w_{ij}h_{ik}(x^n).$$

Since $h_{ik}(x^n) \in \overline{C}$, (1.19) follows.

Identity (1.19) shows that $\chi_\rho(w_{ij}^{-1}c^{-1}w_{ij}c) = 1$ for any $c \in \overline{C}$ if and only if χ_ρ is trivial on $h_{ij}(F^{\times n})$. The lemma then follows. □

Let $\overline{H(B)}$ be the subgroup of $\overline{T}$ generated by $\overline{H_{ij}(B)}$, $i, j \leq r$. The $\overline{H(B)}$ is an abelian group. In this paper, we alway take the maximal abelian subgroup $\overline{A}$ of $\overline{T}$ satisfying the following condition:

$$\overline{H(B)} \subset \overline{A}. \tag{1.20}$$

In particular, $h_{ij}(-1) \in \overline{A}$ for any $i, j \leq r$.

COROLLARY 1.10. *Assume $r \geq 3$ and we consider $\overline{GL(r, F)}$. Let ρ be an irreducible representation of $\overline{T}$. If $w_{ij}(\rho) = \rho$, there is a character ρ° of $\overline{A}$ such that $w_{ij}(\rho^\circ) = \rho^\circ$, $\rho^\circ(H_{ij}(B)) = 1$, and $\rho = i_{\overline{A}}^{\overline{T}}\rho^\circ$.*

PROOF. By Lemma 1.9, χ_ρ is trivial on the group 1.19. We need to show that χ_ρ can be extended to a trivial character ρ° of the group $D = \{w_{ij}^{-1}aw_{ij}a^{-1} : a \in \overline{A}\}$. The lemma then follows from Lemma 1.3. To this end, we observe that by (1.17),

$$D \cap \overline{C} \subset s(H_{ij}) \cap \overline{C} = s(H_{ij}^n).$$

Hence χ_ρ is trivial on $D \cap \overline{C}$ and the desired extension of χ_ρ to $\overline{A}$ can be constructed. □

1.3. Intertwining Operators

Let F be a p-adic field. Denote by $|\cdot|$ the normalized valuation, which is given by $|\varpi| = p^{-f}$, $f = [\mathfrak{O}/\varpi\mathfrak{O} : \mathbf{Z}_p/p\mathbf{Z}_p]$. Fix a measure on $\mathfrak{O}$ such that $\int_{\mathfrak{O}} dx = |\mathfrak{O}/\mathfrak{d}|^{-1/2}$, where $\mathfrak{d}$ is the absolute difference.

Let M and L be a standard Levi of G such that $M \supset L$. If P is a parabolic subgroup, let Δ_P be the modular function. Let (ρ, V) be an admissible representation of $\overline{L}$. Let $N_M = N \cap \overline{M}$. Define the induced representation of $\overline{M}$ to be the space of all functions $f : \overline{M} \to V$ such that

(1) $f(lnm) = \Delta_{LN_M}^{1/2}(l)\rho(l)f(m), \forall l \in \overline{L}, n \in N_M, m \in \overline{M}$;

(2) there is an open subgroup U of $\overline{M}$ such that $f(mu) = f(m), \forall u \in U, m \in \overline{M}$.

The group $\overline{M}$ acts on this space by right translations. Denote this representation by $i_L^M\rho$.

Let N_- be the group of all unipotent lower triangular matrices. For $\sigma \in \overline{M}$ such that $p(\sigma \overline{L} \sigma^{-1})$ is again a standard Levi subgroup of M, define

$$\begin{aligned} N_\sigma &= N \cap \sigma \overline{N_-} \sigma^{-1} \cap \overline{M}, \\ M(\sigma, \rho): & \quad i_{\overline{L}}^{\overline{M}} \rho \to i_{\overline{\sigma(L)}}^{\overline{M}} \sigma(\rho), \\ (M(\sigma, \rho) f)(g) &= \int_{N_\sigma} f(\sigma^{-1} n g) dn, \quad \forall g \in \overline{M}, \end{aligned}$$

whenever the integral converges.

Let $p = (r_1, r_2, \cdots, r_k)$ be a partition of r, i.e., $r = r_1 + r_2 + \cdots + r_k$. Put $r_0 = 0$. Denote $d(p) = d(M) = k$. Let $G(p)$ be the standard Levi corresponding to the partition of p. Suppose $M = G(p)$. Suppose q is another partition of r and $l = d(q)$, $L = G(q)$.

For $\underline{s} \in \mathbf{C}^l$, let $\chi_{\underline{s}}$ be the character of M given by

$$\chi_{\underline{s}}(\mathrm{diag}(g_1, \cdots, g_l) = |\det(g_1)|^{s_1} \cdots |\det(g_l)|^{s_l}.$$

Define

$$M(\sigma, \rho, \underline{s}) = M(\sigma, \rho \chi_{\underline{s}}).$$

Identify $\sigma \in \overline{M}$ with an element in $\mathfrak{S}_r$ via the Bruhat decomposition $\sigma = ntwn' \mapsto w$, $n, n' \in \overline{N}$, $t \in \overline{T}$ and $w \in W$. Let $\ell(\sigma) = \ell(w)$ be the length of w.

LEMMA 1.11. *The operator $M(\sigma, \rho, \underline{s})$ can be continued to a meromorphic function on $\mathbf{C}^{d(L)}$ and satisfies the following functional equation*

$$M(\tau, \sigma\rho, \sigma\underline{s}) M(\sigma, \rho, \underline{s}) = M(\tau\sigma, \rho, \underline{s}), \quad \forall \sigma, \tau \in \overline{M}, \ell(\sigma) + \ell(\tau) = \ell(\tau\sigma).$$

We consider the case when $L = T$. Let ρ be an irreducible representation of $\overline{T}$.

LEMMA 1.12. *Let $f \in i_{\overline{T}}^{\overline{M}} \rho$. Suppose $g = h_{ij}(a) k$ with $a \in F^\times$ and $k \in \overline{K}$. Then there is an integer l such that*

$$(M(w_{ij}(1), \rho) f)(g) \tag{1.21}$$

$$= \int_{a^2 \varpi^{-l+1} \mathfrak{O}} f\left(w_{ij}(-1) n_{ij}(x) g\right) dx \tag{1.22}$$

$$+ \int_{\varpi^l \mathfrak{O}} f\left(h_{ij}(a^{-1} x) k\right) c(x, a) \left|\frac{a}{x}\right|^2 dx. \tag{1.23}$$

PROOF. Let $U_l = \{k \in K, k \cong 1 (\mod \varpi^l \mathfrak{O})\}$. Choose l large enough so that f is invariant under the right multiplication of $s(U_l)$ and $s(U_l)$ is normal in $\overline{K}$. Observe that $F = \varpi^{-l+1} \mathfrak{O} \cup \{y \in F : y^{-1} \in \varpi^l \mathfrak{O}\}$. So

$$(M(w^{ij}(1), \rho) f)(g) = \int_{a^{-2} x \in \varpi^{-l+1} \mathfrak{O}} + \int_{a^2 x^{-1} \in \varpi^l \mathfrak{O}} f\left(w_{ij}(-1) n_{ij}(x) g\right) dx.$$

By (1.3), the last integral is

$$\int_{a^2 x^{-1} \in \varpi^l \mathfrak{O}} f\left(n_{ij}(-x^{-1}) h_{ij}(x^{-1}) h_{ij}(a) n_{ji}(a^2 x^{-1}) k\right) dx.$$

Since $n_{ji}(a^2 x^{-1}) \in s(U_l)$, it can be omitted. The identity in the lemma follows after the change of variable $y = a^2 x^{-1}$. □

We can further simplify (1.23). Write $x = \xi z$ with $\xi \in F^\times/F^{\times n}$ and $z \in F^{\times n}$. Then

$$(1.24) \qquad (1.23) = \sum_{\xi \in F^\times/F^{\times n}} \frac{c(\xi, a)|a|^2}{|\xi|} f\left(h_{ij}(a^{-1}\xi)k\right) \int_{(\xi^{-1}\pi^l\mathfrak{O})\cap F^{\times n}} \rho_{ij}(z)\frac{dz}{|z|}.$$

COROLLARY 1.13. *Let ρ be a unitary irreducible representation of $\overline{T}$. The operator $M(w_{ij}, \rho, \underline{s})$ has a pole at $\underline{0}$ if and only if ρ_{ij}^n is unramified and has exponent 0. If this is the case, the pole is simple.*

PROOF. For simplicity, replace $\rho\chi_{\underline{s}}$ by ρ and consider $M(w_{ij}, \rho)$. Consider the identity in Lemma 1.12. The term (1.22) is analytic since the domain of integration is compact. Now study the meromorphic behavior of (1.23) which is equal to (1.24). Each summand in (1.24) equals

$$(1.25) \qquad \int_{\mathfrak{O}^n} \rho_{ij}(z)\frac{dz}{|z|} = \frac{|n|_F}{n\left(1-\rho_{ij}^n(\varpi)\right)} \int_{\mathfrak{O}^\times} \rho_{ij}^n(z)dz$$

up to some constant. Now the corollary follows. □

From lemma 1.12 and (1.24), it is easy to get

$$(1.26) \quad M(w_{ij}(-1), \rho)f(k) = \int_{\varpi^{-l+1}\mathfrak{O}} f(w_{ij}(1)n_{ij}(x)k)dx + \sum_{\xi \in F^\times/F^{\times n}} |\xi|^{-1}c(\xi,-1)f(h_{ij}(\xi)k) \int_{(\xi^{-1}\varpi^l\mathfrak{O})\cap F^{\times n}} \rho_{ij}(z)\frac{dz}{|z|}.$$

In the rest part of this section, assume the following conditions are satisfied

$$(1.27) \qquad \overline{G} \text{ splits over } K; \quad |n|_F = 1; \quad \text{and the absolute difference is } \mathfrak{O}.$$

Denote be K^* the image of K under a splitting homomorphism. We say an irreducible representation ρ of $\overline{T}$ is unramified if the restriction of ρ on $\overline{T} \cap K^* = \overline{T} \cap s(K)$ is the identity operator. By lemma 1.3, this is the case if and only if the central character of ρ is trivial on $\overline{C} \cap s(K)$. In this case, it is easy to see by the Bruhat decomposition that the space of K^*-fixed vector in $i_{\overline{T}}^{\overline{G}}\rho$ is one dimensional. Realize $\rho = i_{\overline{A}}^{\overline{T}}\rho^\circ$ for a character ρ° of $\overline{A}$ whose restriction to $\overline{A} \cap s(K)$ is trivial. Let $1_\rho \in i_{\overline{A}}^{\overline{T}}\rho^\circ$ such that $1_\rho(a) = 1$ for $a \in \overline{A} \cap s(K)$ and $1_\rho(t) = 0$ for $t \notin \overline{A}$. Denote by v_ρ the vector in $i_{\overline{T}}^{\overline{G}}\rho$ such that $v_\rho(1) = 1_\rho$. For convenience, call v_ρ the canonical K^*-fixed vector.

COROLLARY 1.14. *Under conditions (1.27), suppose ρ is an unramified representation of $\overline{T}$ such that the exponent of ρ_{ij}^n is s. Then $M(w_{ij}(1), \rho)v_\rho = (1 - |\varpi|^{s+1})/(1 - |\varpi|^s) \cdot v_{(ij)\rho}$.*

PROOF. We show that $(M(w_{ij}(1), \rho)v_\rho)(1) = (1-|\varpi|^{s+1})/(1-|\varpi|^s)1_{(ij)\rho}$. By Lemma 1.12 and (1.24), the left hand is

$$(1.28) \qquad \int_{\varpi\mathfrak{O}} f(w_{ij}(-1))dx + \sum_{\xi \in F^\times/F^{\times n}} \frac{f(h_{ij}(\xi))}{|\xi|} \int_{(\xi^{-1}\mathfrak{O})\cap F^{\times n}} \rho_{ij}(z)\frac{dz}{|z|}.$$

Since $M(w_{ij}(1),\rho)v_\rho$ is a multiple of $v_{(ij)\rho}$, the terms after the sum sign will be canceled unless $\xi \in F^{\times n}$. Also by (1.12), $w_{ij}(-1) \in K^*$. Now (1.28) valued at $1_{\overline{T}}$ is

$$\begin{aligned} & |\varpi| \int_{\mathfrak{O}} dx + \sum_{\xi \in \mathfrak{O}^{\times n}} \int_{\mathfrak{O}^n} \rho_{ij}(z) \frac{dz}{|z|} \\ = \ & |\varpi| \int_{\mathfrak{O}} dx + \left(1 - \rho_{ij}^n(\varpi)\right)^{-1} \int_{\mathfrak{O}^\times} dz \\ = \ & \frac{1 - |\varpi|^{s+1}}{1 - |\varpi|^s} \int_{\mathfrak{O}} dx. \end{aligned}$$

Recall the measure of $\mathfrak{O}$ is 1 if the absolute difference is $\mathfrak{O}$. Now the lemma follows. □

1.4. Product of Two Intertwining Operators

LEMMA 1.15. *Let F be a non-archimedean field. As a meromorphic operator,*

$$M\left(w_{ij}(-1), w_{ij}(1)(\rho)\right) M\left(w_{ij}(1), \rho\right)$$
$$= \begin{cases} |n|_F \left(\int_{\mathfrak{O}} dx\right)^2 \cdot |\varpi| \frac{\left(1-|\varpi|^{s-1}\right)\left(1-|\varpi|^{s+1}\right)}{(1-|\varpi|^s)^2} \\ \qquad \text{if } \rho_{ij}^n \text{ is unramified and } \rho_{ij}^n = |\cdot|^s; \\ |n|_F \left(\int_{\mathfrak{O}} dx\right)^2 \cdot |\varpi|^c \\ \qquad \text{if } \rho_{ij}^n \text{ is ramified with conductor } U_c. \end{cases}$$

PROOF. Identify $\overline{C}\backslash\overline{T}$ with a fixed set of representatives of it. Let $f \in i_{\overline{T}}^{\overline{M}}\rho$. By (1.26), for $k \in \overline{C}\backslash\overline{T}K$, there is an l which is a large multiple of n, such that

$$\begin{aligned} (1.29)\quad & M(w_{ij}(-1), w_{ij}(1)\rho)M(w_{ij}(1),\rho)f(k) \\ (1.30)\quad = \ & \int_{\varpi^{-l+1}\mathfrak{O}} (M(w_{ij}(1),\rho)f)(w_{ij}(1)n_{ij}(x)k)dx \\ (1.31)\quad + \ & \sum_{\xi \in F^\times/F^{\times n}} \frac{c(\xi,-1)}{|\xi|}(M(w_{ij},\rho)f)(h_{ij}(\xi)k) \int_{(\xi^{-1}\varpi^l\mathfrak{O})\cap F^{\times n}} \rho_{ij}^{-1}(z)\frac{dz}{|z|}. \end{aligned}$$

Any of the above integrals is understood as a meromorphic function of the exponent of ρ_{ij}^n whose values are given by the integral where it converges. The same understanding applies to the following calculations. Identify the quotient group $F^\times/F^{\times n}$ with a fixed set of representatives in $F^\times$.

We need to show (1.29) is $f(k)$ multiplied by the constant specified in the lemma. Applying (1.4) to (1.30), we get

$$\begin{aligned} (1.30) = & \iint_{x\in\varpi^{-l+1}\mathfrak{O}} \\ & f\left(n_{ij}\left(x(1-xy)^{-1}\right)h_{ij}\left(\left(1-xy^{-1}\right)n_{ji}\left(y(xy-1)^{-1}\right)\right)c(1-xy,-x)dxdy \\ (1.32)\quad & \iint_{1-u\in v\varpi^{-l+1}\mathfrak{O}} f(h_{ij}(u))n_{ji}(v)k)c(u,-v)\frac{dudv}{|u^2v|} \\ (1.33)\quad & \int_{v\in\varpi^l\mathfrak{O}}\int_{1-u\in v\varpi^{-l+1}\mathfrak{O}} f(h_{ij}(u)k)c(u,-v)\frac{dudv}{|u^2v|} \\ (1.34)\quad + & \int_{v\notin\varpi^l\mathfrak{O}}\int_{u\in v\varpi^{-l+1}\mathfrak{O}} f(h_{ij}(u)n_{ji}(v)k)c(u,-v)\frac{dudv}{|u^2v|}. \end{aligned}$$

We used the following change of variables to get (1.32):

$$u=\frac{1}{1-xy},\quad v=\frac{y}{xy-1},\quad \left|\frac{\partial(x,y)}{\partial(u,v)}\right|=\frac{1}{|u^2v|}.$$

In (1.33), replace v by $-v\varpi^l$. It then becomes (recall we assume $n|l$.)

$$\int_{u\in 1+\varpi\mathfrak{O}} f(h_{ij}(u)k)\frac{du}{|u|^2}\int_{\mathfrak{O}-(1-u)\mathfrak{O}} c(u,v)\frac{dv}{|v|}. \tag{1.35}$$

Observe the inner integral is 0 unless $c(u,x)=1$ for any $x\in\mathfrak{O}^\times$. On the other hand, suppose $u\in 1+\varpi^j\mathfrak{O}^\times$ and $c(u,x)=1,\forall x\in\mathfrak{O}^\times$. The identity $c(u,1-u)=1$ then implies that $c(u,\varpi^j)=1$. The inner integral of (1.35) is

$$\sum_{i=0}^{j-1}\int_{\pi^i\mathfrak{O}^\times} c(u,v)\frac{dv}{|v|}=\sum_{i=0}^{j-1}c(u,\pi)^i\int_{v\in\mathfrak{O}^\times}dv.$$

This is 0 unless $c(u,\pi)=1$. Hence we conclude that the inner integral of (1.35) is 0 unless $u\in F^{\times n}$ and (1.33)=(1.35) becomes

$$\sum_{j\geq 1} j\int_{(1+\varpi^j\mathfrak{O}^\times)\cap F^{\times n}}\rho_{ij}(u)du\int_{\mathfrak{O}^\times}dvf(k)=|n|_F|\varpi|^c\left(\int_{\mathfrak{O}}du\right)^2 f(k). \tag{1.36}$$

The proof of the last identity is postponed to lemma 1.16.

Leaving term (1.34) aside, we now calculate (1.31). By Lemma 1.12,

(1.31) =

$$\sum_{\xi\in F^\times/F^{\times n}}\int_{\varpi^l\mathfrak{O}} f(h_{ij}(\xi^{-1}y)k)c(-y,\xi)\frac{|\xi|dy}{|y|^2}\int_{(\xi^{-1}\varpi^l\mathfrak{O})\cap F^{\times n}}\rho_{ij}^{-1}(z)\frac{dz}{|z|} \tag{1.37}$$

$$+\sum_{\xi\in F^\times/F^{\times n}}\frac{c(\xi,-1)}{|\xi|}\int_{\xi^{-2}x\in\varpi^{-l+1}\mathfrak{O}} f(w_{ij}(-1)n_{ij}(x)h_{ij}(\xi)k)dx$$

$$\int_{(\xi^{-1}\varpi^l\mathfrak{O})\cap F^{\times n}}\rho_{ij}^{-1}(z)\frac{dz}{|z|}. \tag{1.38}$$

It is obvious that (1.37) is 0 unless ρ_{ij}^n is unramified. Assume this when we discuss (1.37). Also the ϖ^l can be removed in (1.37) since $n|l$. Now let $y=x\eta$ for $x\in F^{\times n}$ and $\eta\in F^\times/F^{\times n}$. Then (1.37) becomes

$$\sum_{\xi,\eta\in F^\times/F^{\times n}}\left|\frac{\xi}{\eta}\right| f(h_{ij}(\xi^{-1}\eta)k)c(-\eta,\xi)\int_{\eta^{-1}\mathfrak{O}\cap F^{\times n}}\rho_{ij}(x)\frac{dx}{|x|}\int_{\xi^{-1}\mathfrak{O}\cap F^{\times n}}\rho_{ij}^{-1}(z)\frac{dz}{|z|}.$$

Recall the above two integrals are understood as meromorphic functions of the exponent of ρ_{ij}. The η^{-1} and ξ^{-1} in the above two integral domains can be removed if we require $0\leq|\xi|,|\eta|\leq|\varpi|^{n-1}$. Let $\delta=\eta/\xi$ and then (1.37) becomes

$$\sum_{\delta\in F^\times/F^{\times n}}|\delta|^{-1}f(h_{ij}(\delta)k)\sum_{\xi\in F^\times/F^{\times n}}c(\delta,\xi)\int_{\mathfrak{O}^n}\rho_{ij}^{-1}(x)\frac{dx}{|x|}\int_{\mathfrak{O}^n}\rho_{ij}(x)\frac{dx}{|x|}$$

$$= -|n|_F\rho_{ij}^n(\varpi)\left(\frac{\int_{\mathfrak{O}^\times}\rho_{ij}^n(x)\,dx}{1-\rho_{ij}^n(\varpi)}\right)^2 f(k). \tag{1.39}$$

In the last step, we used the fact that the second sum is 0 unless $\delta\in F^{\times n}$. Observe that (1.39) is true even when ρ_{ij}^n is ramified.

So (1.29) is the sum of (1.34), (1.36), (1.38) and (1.39). Now the lemma follows by a straightforward calculation if we can show that (1.34) + (1.38) = 0. To see this cancellation, first plug the identity

$$h_{ij}(u)n_{ji}(v)c(u,-v) = h_{ij}(uv^{-1})w_{ij}(-1)n_{ij}(v^{-1})c(uv^{-1},-1)$$

into (1.34), then make the following change of variables:

$$x = v^{-1},\ y = uv^{-1},\ \frac{\partial(u,v)}{\partial(x,y)} = \frac{1}{x^3}.$$

Expression (1.34) now becomes

$$\begin{aligned}&\int_{\varpi^{-l+1}\mathfrak{O}}\int_{\varpi^{-l+1}\mathfrak{O}} f(h_{ij}(y)w_{ij}(-1)n_{ij}(x)k)c(y,-1)\frac{dxdy}{|y|^2}\\ =\ &\sum_{\xi\in F^\times/F^{\times n}} \frac{c(\xi,-1)}{|\xi|}\int_{\varpi^{-l+1}\mathfrak{O}} f(h_{ij}(\xi)w_{ij}(-1)n_{ij}(x)k)dx\\ &\int_{(\xi^{-1}\varpi^{-l+1}\mathfrak{O})\cap F^{\times n}} \rho_{ij}(z)\frac{dz}{|z|}.\end{aligned}$$

We replaced y by ξz, for $z \in F^{\times n}$ and $\xi \in F^\times/F^{\times n}$, to get the last identity. Make the following changes of variables $x \mapsto \xi^2 x$ and then $\xi \mapsto \xi^{-1}$ in (1.38). Now the identity (1.34) + (1.38) = 0 is an immediate consequence of

$$\int_{(\xi^{-1}\varpi^{-l+1}\mathfrak{O})\cap F^{\times n}} \rho_{ij}(z)\frac{dz}{|z|} + \int_{(\xi\varpi^{l}\mathfrak{O})\cap F^{\times n}} \rho_{ij}^{-1}(z)\frac{dz}{|z|} = 0$$

which can be easily deduced from (1.25). Remark again that the above integrals are meromorphic functions of the exponent of ρ_{ij}^n. □

Now we need to prove (1.36). It follows from the following lemma.

LEMMA 1.16. *Let ν be a character of $F^{\times n}$ and c the least positive integer such that ν^n is trivial on $1+\varpi^c\mathfrak{O}$. Then*

$$\sum_{j\geq 1} j\int_{(1+\varpi^j\mathfrak{O}^\times)\cap F^{\times n}} \nu(u)du = \frac{|n|_F|\varpi|^c}{1-|\varpi|}\int_{\mathfrak{O}} du.$$

PROOF. We show by induction on $|n|_F$ that

$$\sum_{j\geq 1}\int_{(1+\varpi^j\mathfrak{O})\cap F^{\times n}} \nu(u)du = \frac{|n|_F|\varpi|^c}{1-|\varpi|}\int_{\mathfrak{O}} du \tag{1.40}$$

from which the lemma will follow.

First suppose that $|n|_F = 1$, then $1+\varpi\mathfrak{O} \subset F^{\times n}$ and c is a also the smallest positive integer such that ν is trivial on $1+\varpi^c\mathfrak{O}$. Hence when $j<c$, the integral on the left hand side of (1.40) vanishes and the proof is straightforward.

Suppose it is true for n. We calculate for np. Let $U_i = 1+\pi^i\mathfrak{O}$ for $i\geq 1$. Let e be the ramification index for $F/\mathbf{Q}_p$, i.e., $p = \varpi^e\mathfrak{O}^\times$. First observe that if F contains the p-th roots of unity, then

$$U_i = (U_{i-e})^p, \qquad \text{if } i > \frac{ep}{p-1}; \tag{1.41}$$

$$U_{ip-j}\cap F^{\times p} = (U_i)^p, \qquad \text{if } i \leq \frac{e}{p-1}, 0\leq j<p. \tag{1.42}$$

By this observation, If $i > pe/(p-1)$,

$$\int_{(1+\varpi^i\mathfrak{O})\cap F^{\times np}} \nu(u)\,du = \int_{(U_{i-e})^p\cap F^{\times np}} \nu(u)\,du = |p| \int_{U_{i-e}\cap F^{\times n}} \nu(v^p)\,dv.$$

For the last identity, we changed variables: $u = v^p$. (Notice that U_{i-e} does not contain any primitive p-th root of unity.) If $i \le \frac{e}{p-1}$, $0 \le j < p$,

$$\int_{U_{ip-j}\cap F^{\times np}} \nu(u)\,du = \int_{(U_i)^p\cap F^{\times np}} \nu(u)\,du = \frac{|p|}{p} \int_{U_i\cap F^{\times n}} \nu(v^p)\,dv.$$

Again we used the change of variable $u = v^p$. (In this case, U_i contains all p-th roots of unity.) So

$$\begin{aligned} &\sum_{i=1}^{\infty} \int_{U_i\cap F^{\times np}} \nu(u)\,du \\ = &\ |p| \sum_{i>e/(p-1)} \int_{U_i\cap F^{\times n}} \nu(v^p)\,dv + |p| \sum_{i\le e/(p-1)} \int_{U_i\cap F^{\times n}} \nu(v^p)\,dv \\ = &\ |p| \sum_{i\ge 1} \int_{U_i\cap F^{\times n}} \nu^p(v)\,dv. \end{aligned}$$

Now the assertion (1.40) for np follows from the induction hypothesis. □

1.5. Normalization of Intertwining Operators

Let F be an local field. From now on we suppose that $c(-1,-1) = 1$. Recall that we have fixed a maximal abelian subgroup $\overline{A}$ of $\overline{T}$ satisfying (1.20).

Assume that F is a local field. Fix an additive character of F once for all. Let ν be a quasicharacter of $F^\times$. Let $L(s,\nu)$ and $\varepsilon(s,\nu)$ be the L-function and the ε-factor respectively. Recall the following definitions.

for non-archimedean field
  if ν is ramified
    $L(s,\nu) = 1$;
  if ν is unramified and $\nu = |\cdot|^{s_0}$
    $L(s,\nu) = (1 - |\varpi|^{s+s_0})^{-1}$, $\varepsilon(x,\nu) = \left(\int_{\mathfrak{O}} dx\right)^{2(s+s_0)-1}$;
for the complex field
  if $\nu(x) = |x|^{s_0}x^{-m}$ or $|x|^{s_0}\overline{x}^{-m}$ $(m \ge 0)$
    $L(s,\nu) = (2\pi)^{1-s-s_0}\,\Gamma(s+s_0)$;

If F is p-adic, $\varepsilon(s,\nu)$ is an entire function satisfying the following relation

$$\varepsilon(s,\nu)\,\varepsilon(-s,\nu^{-1}) = \nu(-1)|\varpi|^{-c}\left(\int_{\mathfrak{O}} dx\right)^{-2} \tag{1.43}$$

where c is the least nonnegative integer such that ν is trivial on $1 + \varpi^c\mathfrak{O}$.

For an irreducible representation ρ of $\overline{T}$ and $\underline{s} \in \mathbf{C}^r$, fix a character ρ° of $\overline{A}$ which extends the central character of ρ. For $i,j \le r$, let $\gamma(\rho_{ij}) = 1$ if $\rho^\circ(h_{ij}(-1))^{n-1} = 1$ and $\gamma(\rho_{ij}) = \mathbf{i}$ if $\rho^\circ(h_{ij}(-1))^{n-1} = -1$. Here $\mathbf{i}$ is fixed primitive fourth root of unity. By definition, $\gamma(\rho_{ij})^2 = \rho^\circ(h_{ij}(-1))^{n-1}$. In particular,

$\gamma(\rho_{ij}) = 1$ if n is odd. Define for $\sigma \in \mathfrak{S}_r$,

$$
\begin{aligned}
(1.44) \qquad \operatorname{inv}(\sigma) &= \{i, j \in \mathbf{Z} | 1 \le i < j \le r, \sigma(i) > \sigma(j)\}; \\
(1.45) \qquad M(\sigma, \rho, \underline{s}) &= M(w_\sigma, \rho, \underline{s}); \\
(1.46) \qquad r(\sigma, \rho, \underline{s}) &= \prod_{i,j \in \operatorname{inv}(\sigma)} \frac{\gamma(\rho_{ij})|n|_F^{1/2} L\left(n(s_i - s_j), \rho_{ij}^n\right)}{L\left(n(s_i - s_j) + 1, \rho_{ij}^n\right) \varepsilon\left(n(s_i - s_j), \rho_{ij}^n\right)}; \\
(1.47) \qquad N(\sigma, \rho, \underline{s}) &= r(\sigma, \rho, \underline{s})^{-1} M(\sigma, \rho, \underline{s}).
\end{aligned}
$$

PROPOSITION 1.17. *The operator $N(\sigma, \rho, \underline{s})$ is normalized:*
(1) $N(\tau\sigma, \rho, \underline{s}) = N(\tau, \sigma\rho, \sigma(\underline{s})) N(\sigma, \rho, \underline{s})$.
(2) If F is non-archimedean, $N(\sigma, \rho, \underline{s})$ is analytic in the domain

$$\{\underline{s} \in \mathbf{C}^r : \rho_{ij}^n| \cdot |^{n(s_i - s_j)} \ne | \cdot |^{-1}, \forall (i,j) \in \operatorname{inv}(\sigma)\}.$$

If $F = \mathbf{C}$, $N(\sigma, \rho, \underline{s})$ is analytic in the domain

$$\{\underline{s} \in \mathbf{C}^r : \rho_{ij}| \cdot |^{s_i - s_j} \notin | \cdot |^{-\mathbf{N}}, \forall (i,j) \in \operatorname{inv}(\sigma)\}.$$

(3) Assume that conditions (1.27) are satisfied. If ρ is unramified, then

$$N(\sigma, \rho, \underline{s}) v_\rho = v_{\sigma\rho}.$$

(4) It is unitary when ρ is unitary and $\underline{s} \in \mathbf{iR}^{r-1}$.

PROOF. (1) The identity holds when $\ell(\tau\sigma) = \ell(\tau) + \ell(\sigma)$. So we only need to show that if $\sigma = (i, i+1)$,

$$N(\sigma, \sigma\rho, \sigma\underline{s}) N(\sigma, \rho, \underline{s}) = 1. \tag{1.48}$$

Observe that

$$N(\sigma, \sigma\rho, \sigma\underline{s}) = \rho_{i,i+1}(-1) N(w_\sigma(-1), \sigma\rho, \sigma\underline{s}).$$

By lemma 1.15, the left hand side of (1.48) is

$$
\begin{aligned}
& r(\sigma, \sigma\rho, \sigma\underline{s})^{-1} r(\sigma, \rho, \underline{s})^{-1} M(\sigma, \sigma\rho, \sigma\underline{s}) M(\sigma, \rho, \underline{s}) \\
= \ & \gamma((\sigma\rho)_{i,i+1})^{-1} \gamma(\rho_{i,i+1})^{-1} \rho_{i,i+1}(-1)^{n-1}.
\end{aligned}
$$

By the definition of $\gamma(\rho_{i,i+1})$, the last expression is 1.
(2) follows from Corollary 1.13 and basic properties of the L-functions.
(3) follows from Corollary 1.14.
(4) follows from (1). □

1.6. The Special Value of an Intertwining Operator

Assume F is p-adic. The result is this section is not used for $\overline{GL(2)}$. So assume $r \ge 3$. Keep the notation in the previous section. Fix $i, j \le r$. Let $w = w_{ij} = w_{ij}(1)$. If $\underline{s} \in \mathbf{C}^r$, let $s = s_i - s_j$. Denote by $\mathbf{i}$ a square root of -1. Assume ρ_{ij}° is unramified. By Corollary 1.10, we can extend ρ to a character $\rho_\circ$ of $\overline{A}$ such that $\rho^\circ(h_{ij}(-1)) = 1$.

PROPOSITION 1.18. *If $w(\rho) = \rho$, as a meromorphic operator,*

$$\lim_{s \to 0} N(w, \rho, \underline{s}) = 1. \tag{1.49}$$

PROOF. If F is the field of complex numbers, it is a special case of [**KS88**, Proposition 6.3]. Now assume F is a p-adic field. In this case,

$$L(ns,1) = (1-|\varpi|^{ns})^{-1}.$$

By Corollary 1.13, the operator $\lim_{s\to 0}(1-|\varpi|^{ns})\,M(w,\rho,\underline{s})$ is bounded. Now apply Lemma 1.12. The domain of the integral in (1.22) is compact and hence the integral can be ignored when applying limit. By (1.24) with $a=1$, when $s\to 0$, for $f \in i_{\overline{T}}^{\overline{M}}\rho$, $k\in\overline{K}$,

$$\begin{aligned}
&(1-|\varpi|^{ns})\,M(w,\rho,\underline{s})\,f(k)\\
&\sim\ (1-|\varpi|^{ns})\sum_{\xi\in F^\times/F^{\times n}}\frac{f(h_{ij}(\xi)k)}{|\xi|}\int_{(\xi^{-1}\varpi^l\mathfrak{O})\cap F^{\times n}}|z|^s\frac{dz}{|z|}\\
&\sim\ \sum_{\xi\in F^\times/F^{\times n}}\frac{f(h_{ij}(\xi)k)}{|\xi|}(1-|\varpi|^{ns})\int_{\mathfrak{O}^{\times n}}|z|^s\frac{dz}{|z|}
\end{aligned}$$

Write $\xi=\epsilon\eta$ with $\epsilon\in F^\times/B$ and $\eta\in B/F^{\times n}$. Apply (1.25) to the last integral. Then the above expression becomes

$$\begin{aligned}
&=\ \sum_{\epsilon\in F^\times/B}\sum_{\eta\in B/F^{\times n}}\frac{f(h_{ij}(\eta)h_{ij}(\epsilon)k)}{|\eta\epsilon|}c(\eta,\epsilon)^{-1}\left(\frac{|n|_F}{n}\int_{\mathfrak{O}^\times}dz\right)\\
&=\ \sum_{\eta\in F^\times/B}\frac{f(h_{ij}(\epsilon)k)}{|\xi|}\left(\sum_{\eta\in B/F^{\times n}}c(\epsilon,\eta)\right)\left(\frac{|n|_F}{n}\int_{\mathfrak{O}^\times}dz\right)
\end{aligned}$$

The inner sum is 0 unless $\epsilon\in B$. So the above calculation shows that

$$(1.50)\qquad (1-|\varpi|^{ns})\,M(w,\rho,\underline{s})\,f(k)\sim\frac{f(k)|B/F^{\times n}||n|_F}{n}\int_{\mathfrak{O}^\times}dz$$

Notice in this case, $\gamma(\rho_{ij})=1$. The lemma follows from (1.50) and Lemma 1.5. □

1.7. The Langlands Quotient

Let F be p-adic. It is not hard to see the existence and uniqueness of the Langlands quotient in metaplectic case. The proof is identical with the nonmetaplectic case. As before, let ρ be an irreducible representation of $\overline{T}$. Let w_0 be the longest element in $\mathfrak{S}_r$. for a representation π of $\overline{G}$, denote by $\mathrm{JH}(\pi)$ the set with multiplicities of all composition factors in the Jordan-Hölder series of π.

PROPOSITION 1.19. *Suppose $\rho_{i,i+1}^n$ has exponent no less that zero for any $i<r$. Then the image $L(\rho)$ of $M(w_0,\rho): i_{\overline{A}}^{\overline{G}}\rho\to i_{\overline{A}}^{\overline{G}}w_0(\rho)$ is irreducible. Furthermore, it is the unique irreducible quotient of $i_{\overline{A}}^{\overline{G}}\rho$ and the unique irreducible subrepresentation of $i_{\overline{A}}^{\overline{G}}w_0(\rho)$. And $L(\rho)$ occurs in $\mathrm{JH}(i_{\overline{A}}^{\overline{G}}\rho)=\mathrm{JH}(i_{\overline{A}}^{\overline{G}}w_0(\rho))$ with multiplicity one.* □

As a corollary to the above proposition, we can prove

LEMMA 1.20. *Let ρ be an irreducible representation of $\overline{T}$ such that for any $i<j$, $\rho_{ij}^n\neq|\cdot|$ and the exponent of the character ρ_{ij}^n is not less than zero. Then $i_{\overline{T}}^{\overline{G}}\rho$ is irreducible.*

PROOF. Let w_0 be the longest element is $\mathfrak{S}_r$. Then $N(w_0, w_0\rho)N(w_0,\rho) = 1$ by Proposition 1.17. By the above proposition, $N(w_0, \rho)$ maps the unique irreducible quotient of $i_{\overline{T}}^{\overline{G}}\rho$ to the unique irreducible subrepresentation of $i_{\overline{T}}^{\overline{G}}w_0(\rho)$. Since both $N(w_0, \rho, \underline{s})$ and $N(w_0, w_0\rho, w_0\underline{s})$ are analytic at $\underline{s} = \underline{0}$, $N(w_0, w_0\rho)$ is an isomorphism from $i_{\overline{T}}^{\overline{G}}w_0(\rho)$ to $i_{\overline{T}}^{\overline{G}}\rho$. In particular, $N(w_0, w_0\rho)$ sends the unique irreducible subrepresentation of $i_{\overline{T}}^{\overline{G}}w_0(\rho)$ to the (unique) irreducible subrepresentation of $i_{\overline{T}}^{\overline{G}}\rho$. Summing up, $N(w_0, w_0\rho)N(w_0,\rho) = 1$ maps the unique irreducible subquotient of $i_{\overline{T}}^{\overline{G}}\rho$ to the unique irreducible subrepresentation of $i_{\overline{T}}^{\overline{G}}\rho$. By the multiplicity one for the Langlands quotient, this can happen only if $i_{\overline{T}}^{\overline{G}}\rho$ is irreducible. □

PROPOSITION 1.21. *Let ρ be an irreducible representation of $\overline{T}$ such that for any i, j, $\rho_{ij}^n \neq |\cdot|^{\pm 1}$. Then $i_{\overline{T}}^{\overline{G}}\rho$ is irreducible.*

PROOF. Let w be an element in $\mathfrak{S}_r$ such that the exponent of the character $(w\rho)_{ij}^n$ is no less than zero. Then $i_{\overline{T}}^{\overline{G}}w(\rho)$ is irreducible by the above lemma. Write $w = w_1 w_2 \cdots w_k$ as a product of simple transposition. Then we have

$$N(w,\rho) = N(w_1,\rho)\cdots N(w_k,\rho) : i_{\overline{T}}^{\overline{G}}\rho \to i_{\overline{T}}^{\overline{G}}w(\rho).$$

Observe that each operator on the right hand side of the identity is an isomorphism. So $i_{\overline{T}}^{\overline{G}}\rho \cong i_{\overline{T}}^{\overline{G}}w(\rho)$ is irreducible. □

Let ρ is an irreducible representation of $\overline{T}$ and w is an element in $\mathfrak{S}_r$ such that for any $i, j \leq r$, the exponent of $(w\rho)_{ij}^n$ is no less than 0. Denote by $L(\rho)$ the unique irreducible quotient of $i_{\overline{T}}^{\overline{G}}w\rho$. The following lemma follows from the general theory of representations of locally compact groups.

LEMMA 1.22. *Let $\check{\rho}$ be the contragredient of ρ. Then the contragredient of $L(\rho)$ is $L(\check{\rho})$.* □

Let $p = (r_1, r_2, \cdots, r_k)$ be a partition of r, i.e., $r = r_1 + r_2 + \cdots + r_k$. Put $r_0 = 0$. Let M be the standard Levi corresponding to the partition of p. Let ρ be an irreducible representation of $\overline{A}$ such that for any $i \leq k$ and any j, $r_{i-1}+1 \leq j < r_i$, the exponent of $\rho_{j,j+1}^n$ is no less than 0. Then $i_{\overline{T}}^{\overline{M}}\rho$ has a unique irreducible quotient, denoted π.

We identify an element $\underline{s} \in \mathbf{C}^k$ with $(s_1, s_1, \cdots, s_1, s_2, \cdots, s_k) \in \mathbf{C}^r$ where each s_i occurs r_i times. Identify an element $\sigma \in \mathfrak{S}_k$ with an element in $\mathfrak{S}_r$ permuting intervals, i.e.,

$$\sigma(r_1 + \cdots + r_{i-1} + j) = r_{\sigma^{-1}(1)} + r_{\sigma^{-1}(2)} + \cdots + r_{\sigma^{-1}(i-1)} + j, 1 \leq i \leq k, 1 \leq j \leq r_i.$$

Define an intertwining operator

$$\Upsilon_M : i_{\overline{T}}^{\overline{G}}\rho \to i_{\overline{M}}^{\overline{G}} i_{\overline{T}}^{\overline{M}}\rho \qquad (\Upsilon_M f)(g)(m) = \Delta_{MN}(m)^{-1/2} f(mg)$$

for any $f \in i_{\overline{T}}^{\overline{G}}\rho$ and $g \in \overline{G}$, $m \in \overline{M}$. It is easy to see that Υ provides an isomorphism with the inverse given by $f \mapsto f(g)(1)$. Furthermore, we have

$$M(\sigma, i_{\overline{T}}^{\overline{M}}\rho, \underline{s})\Upsilon_M = \Upsilon_{\sigma M} M(\sigma, \rho, \underline{s}).$$

The following lemma is a corollary of Proposition of 1.17 and the above observation.

LEMMA 1.23. *Notation as above*

1) $N(\tau\sigma, \pi, \underline{s}) = N(\tau, \sigma\pi, \sigma\underline{s})\, N(\sigma, \pi, \underline{s})\,, \ \forall \sigma, \tau \in \mathfrak{S}_k$.

2) $N(\sigma, \pi, \underline{s})$ *is holomorphic at* $\underline{s}$ *if* $N(\sigma, \rho, \underline{s})$ *is.*

3) If conditions (1.27) are satisfied, then $N(\sigma, \pi, \underline{s})\, v_\pi = v_{\sigma\pi}$.

4) $N(\sigma, \pi, \underline{s})$ *is unitary when* $\underline{s} \in i\mathbf{R}^k$ *and* π *is unitary.* □

We close this chapter by remarks on notation. So far all the statements on induced representations and intertwining operators are within $\overline{GL(r)}$. But it is not hard to see that they are also true for a standard Levi subgroup. For example, in the remaining part, fix the standard Levi $G(p)$ corresponding to a partition p of r. Proposition 1.21 can also be stated as

> *Let* ρ *be an irreducible representation of* $\overline{T}$ *such that for any* i, j, $\rho^n_{ij} \neq |\cdot|^{\pm 1}$. *Then* $i_{\overline{T}}^{\overline{G(p)}}\rho$ *is irreducible.*

Suppose ρ is an irreducible representation of $\overline{T}$ such that the exponent of $\rho^n_{i,i+1}$ is not less than 0 for any $i \notin \{r_1 + \cdots + r_j : 1 \leq j \leq k\}$. The corresponding statement to Proposition 1.19 is

> *The induced representation* $i_{\overline{T}}^{\overline{G(p)}}\rho$ *is has a unique irreducible quotient, denoted* $L(\rho, p)$, *which occurs in* $i_{\overline{T}}^{\overline{G(p)}}\rho$ *with multiplicity one.*

Let $\sigma \in \mathfrak{S}_k$. As before, we also identify it with an element in $\mathfrak{S}_r$ which permutes intervals. Then we have

LEMMA 1.24.

$$L(\sigma\rho, \sigma p) = \sigma L(\rho, p).$$

PROOF. As representations of $\overline{\sigma(G(p))}$,

$$i_{\overline{T}}^{\overline{G(\sigma p)}}\sigma\rho = \sigma\left(i_{\overline{T}}^{\overline{G(p)}}\rho\right).$$

Denote by $Q(\pi)$ the irreducible quotient of π for a representation π with a unique irreducible quotient. According to the above identity, we have

$$Q\left(i_{\overline{T}}^{\overline{G(\sigma p)}}\sigma\rho\right) = Q\left(\sigma\left(i_{\overline{T}}^{\overline{G(p)}}\rho\right)\right) = \sigma\left(Q\left(i_{\overline{T}}^{\overline{G(p)}}\rho\right)\right).$$

The lemma then follows. □

CHAPTER 2

Local Intertwining Operators

2.1. Irreducibility and Intertwining Operators

Assume F is a p-adic field in this section. Corresponding to each partition $p = (p_1, \cdots, p_k)$ of r, there is a standard Levi subgroup $G(p) = GL(p_l) \times \cdots \times GL(p_k)$. Denote by $\mathfrak{S}(p)$ the subgroup of $\mathfrak{S}_r$ consisting of all the elements which stabilize the intervals corresponding to the partition p. That is

$$\mathfrak{S}(p) = \{\sigma \in \mathfrak{S}_r : p_{i-1} < \sigma(j) \leq p_i, \forall 0 < i \leq k, p_{i-1} < j \leq p_i\}.$$

Fix an irreducible unitary representation δ of $\overline{T}$. Let $\underline{s} \in \mathbf{C}^r$ and $w \in \mathfrak{S}(p)$ such that for any $l \leq k$, $p_{l-1} < i < j \leq p_l$, we have $s_{w^{-1}(i)} \geq s_{w^{-1}(j)}$. Then $i_{\overline{T}}^{\overline{G(p)}} w(\delta[\underline{s}])$ has a unique irreducible quotient, which is denoted by $L(\delta, p, \underline{s})$ or $L(p, \underline{s})$ if the δ is clear from the context.

Let $p = (m, m')$ be a partition of r. Let $\nu \in \mathbf{R}^m, \nu' \in \mathbf{R}^{m'}$ such that

$$\nu_1 \geq \nu_2 \geq \cdots \nu_m, \nu_1' \geq \nu_2' \geq \cdots \nu_{m'}'.$$

Denote $(\nu, \nu') = (\nu_1, \cdots \nu_m, \nu_1' \cdots \nu_{m'}')$. The proof of the following lemma is similar to that in the linear group case.

LEMMA 2.1. (refer to [**MW89**, p. 609])
Let $\underline{s} = (s, s') \in \mathbf{C}^2$, $\Pi = L(\delta, p, (\nu, \nu'))$. *If* $\nu_m + \mathrm{Re}(s) \geq \nu_m' + \mathrm{Re}(s')$, *then*
(1) $N(\sigma, \Pi, \cdot)$ *is holomorphic at* $\underline{s}$ *and* $N(\sigma, \Pi, \cdot) \neq 0$;
(2) $N(\sigma, \sigma\Pi, \cdot)$ *is holomorphic at* $\sigma\underline{s}$ *if and only if* $i_{\overline{G(p)}}^{\overline{G}}\Pi$ *is irreducible;*
(3) if $N(\sigma, \sigma\Pi, \cdot)$ *is holomorphic at* $\sigma\underline{s}$, *then* $N(\sigma, \sigma\Pi, \sigma\underline{s}) \neq 0$. □

If $a, b \in \mathbf{R}$ such that $n(a-b) \in \mathbf{N}$. We call

$$[a, b] = (a, a + 1/n, a + 2/n, \cdots, b - 1/n, b) \tag{2.1}$$

a segment. Two segments $[a, b]$ and $[a', b']$ are called linked if their union is again a segment and neither contains the other.

If $[a, b]$ is a segment and $r = m \times 1 + (n(b-a)+1) + m' \times 1$. Let $p = (1, \cdots, 1, n(b-a) + 1, 1, \cdots, 1)$ be the corresponding partition of r. Assume $(ij)(\delta) = \delta$ for any $m < i, j \leq r - m'$. Identify $[a, b]$ with the element

$$(\underbrace{0, \cdots, 0}_{m}, [a, b], \underbrace{0, \cdots, 0}_{m'}) \in \mathbf{C}^r.$$

Denote $J(a, b) = L(\delta, p, [a, b])$. If $a > b$, set $J(a, b) = \emptyset$.

Suppose $[a,b]$ and $[a',b']$ are two segments such that $n(b-a+b'-a')+2=r$. Define the following partitions of $m+r+m'$:

$$\begin{aligned} p_0 &= (m, \underbrace{1,\cdots,1}_{r}, m'); \\ p &= (m, n(b-a)+1, n(b'-a')+1, m'); \\ q &= (m, n(b-a+b'-a')+2, m'). \end{aligned}$$

Denote

$$([a,b],[a',b']) = (\overbrace{0,\cdots,0}^{m}, a, a+\frac{1}{n},\cdots,b-\frac{1}{n}, b, a', a'+\frac{1}{n},\cdots,b',\overbrace{0,\cdots,0}^{m'}).$$

Let δ be an irreducible representation of $\overline{G(p_0)}$. Define

$$J(a,b)\times_\delta J(a',b') = i_{\overline{G(p)}}^{\overline{G}} L(\delta, p, ([a,b],[a',b'])). \tag{2.2}$$

This is a representation of $\overline{G}=\overline{G(q)}$.

The above notation are too cumbersome. So in a situation as above, we just use the following notation. Let $G=GL(r)$ and T the group of diagonal matrices. Let $p=(n(b-a)+1, n(b'-a')+1)$ and define $J(a,b)\times_\delta J(a',b')$ as (2.2). Even though we are doing everything within $\overline{G}=\overline{GL(r)}$, we still keep in mind that $\overline{G}$ is a really $\overline{M}$ for some standard Levi subgroup in $\overline{GL(R)}$ for $R\geq r$.

Furthermore, if the context is clear, we write

$$J(a,b)\times J(a',b') = J(a,b)\times_\delta J(a',b').$$

In particular, when $a=b$, by convention

$$[a]\times J(a',b') = J(a,a)\times J(a',b').$$

Let $\sigma\in\mathfrak{S}_r$ defined by $\sigma(i)=i+n(b'-a')+1$ for $i\leq n(b-a)+1$ and $\sigma(j)=j+n(b-a)+1$ for $j\leq n(b'-a')+1$. Then σ represents the nontrivial element in $\mathfrak{S}_r/\mathfrak{S}(p)$. We also identify σ with the nontrivial element in $\mathfrak{S}_2$ in an obvious way.

LEMMA 2.2. (refer to [**MW89**, p. 612])

Suppose that $[a,b]$, $[a',b']$ are segments.

(1) If $a+b\geq a'+b'$, then the representation $J(a,b)\times J(a',b')$ admits a unique irreducible quotient which is isomorphic to $L(\delta,p,([a,b],[a',b']))$.

Similarly, $i_{\overline{G(\sigma p)}}^{\overline{G}} L(\sigma\delta,\sigma p,([a',b'],[a,b]))$ admits a unique irreducible submodule which is isomorphic to the same representation $L(\delta,p,([a,b],[a',b']))$.

(2) The representation $\pi = J(a,b)\times J(a',b')$ is irreducible if and only if $N(\sigma,\pi,\underline{s})$ and $N(\sigma,\sigma\pi,\sigma\underline{s})$ are defined at $\underline{s}=(0,0)$. In this case, $\pi\cong L(\delta,p:([a,b],[a',b']))$.

(3) If $n(b-b')\notin\mathbf{Z}$ or $\sigma\delta\neq\delta$, the equivalent conditions in (2) are satisfied.

PROOF. Denote $\underline{a}=(b,b-1/n,\cdots,a+1/n,a,b',b'-1/n,\cdots,a'+1/n,a')$. Let w_0 be the longest element in $\mathfrak{S}(p)$.

By a standard module dominating $J(a,b)\times_\delta J(a',b')$ we mean an induced representation of the form $i_{\overline{T}}^{\overline{G}}\delta'[\underline{z}]$, where $\underline{z}=(z_1,\cdots,z_r)\in\mathbf{C}^r$, such that $\delta'=w(\delta)$ and $\underline{z}=w\underline{a}$ with $\mathrm{Re}z_1\geq\cdots\geq\mathrm{Re}z_r$ for some $w\in\mathfrak{S}_r$.

Fix a standard module, denoted by C^+, dominating $J(a,b)\times J(a',b')$. Fix a $w\in\mathfrak{S}_r$ associated to C^+.

We identify an element $\underline{s} = (s, s')$ of $\mathbb{C}^2$ with the following element in $\mathbb{C}^r$:

$$(s, \cdots, s, s', \cdots, s')$$

where s occurs $n(b-a)+1$ times and s' occurs $n(b'-a')+1$ times. Put

$$C^+(\underline{s}) = I(w\delta[\underline{a}], w\underline{s}).$$

We have the following normalized intertwining operators:

$$C^+(\underline{s}) \xrightarrow{N\left(w^{-1}, w(\delta[\underline{a}]), w\underline{s}\right)} \delta[\underline{a}+\underline{s}] \xrightarrow{N(w_0, \delta[\underline{a}], \underline{s})} \delta[w_0(\underline{a})+\underline{s}].$$

We have

$$\mathrm{Im} N(w_0, \delta[\underline{a}], \underline{s}) = J(a,b)[s] \times J(a',b')[s'].$$

By convention, $J(a,b)[s] \times J(a',b')[s'] = J(a+s, b+s) \times J(a'+s', b'+s')$.

If $n(b-b') \notin \mathbf{Z}$ or $\sigma\delta \neq \delta$, then $N\left(w^{-1}, w(\delta[\underline{a}], w\underline{s})\right)$ is an isomorphism at $\underline{s} = \underline{0}$ by Proposition 1.17. Since C^+ has a unique irreducible quotient $L = L(\delta[\underline{a}])$, C^+ shares a common irreducible quotient with $J(a,b) \times J(a',b')$.

The contragredient of $J(a',b') \times_{\sigma\delta} J(a,b) = i_{\overline{G(\sigma p)}}^{\overline{G}} L(\sigma\delta, \sigma p, ([a',b'],[a,b]))$ is

$$J(-b',-a') \times_{\sigma\check{\delta}} J(-b,-a) = i_{\overline{G(\sigma p)}}^{\overline{G}} L(\sigma\check{\delta}, \sigma p, ([-b',-a'],[-b,-a])),$$

where $\check{\delta}$ is the contragredient of δ. The $\check{\delta}$ is determined by its central character which is the complex conjugate of the central character of δ. The induced module $J(-b',-a') \times_{\sigma\check{\delta}} J(-b,-a)$ has an irreducible quotient $L\left(\check{\delta}[-\underline{a}]\right)$. We see that $L = L(\delta[\underline{a}])$, the contragredient of $L\left(\check{\delta}[-\underline{a}]\right)$, is a subrepresentation of $J(\delta, a', b') \times J(\delta, a, b)$.

Similarly we see that $J(\delta', a', b') \times_{\sigma\delta} J(\delta, a, b)$ admits L as a quotient, and $J(\delta, a, b) \times_\delta J(\delta', a', b')$ admits L as a subrepresentation.

Now

$$\begin{array}{ccc} C^+ & \longrightarrow & J(\delta,a,b) \times_\delta J(\delta',a',b') \\ & \searrow & \updownarrow \\ & & J(\delta',a',b') \times_{\sigma\delta} J(\delta,a,b) \end{array}$$

All operators are normalized intertwining operators. The vertical arrows are isomorphisms by Lemma 1.23, so $J(\delta,a,b) \times_\delta J(\delta',a',b')$ has L as both a submodule and a quotient module. But L has multiplicity one in all subquotients of C^+ by Proposition 1.19, So $J(\delta,a,b) \times J(\delta',a',b')$ must be irreducible.

Now suppose $n(b-b') \in \mathbf{Z}$ and $\sigma\delta = \delta$ in the rest of the proof. Then we have $J(a',b') \times_{\sigma\delta} J(a,b) = J(a',b') \times_\delta J(a,b)$. So we can omit δ. We prove the surjectivity

$$C^+ \twoheadrightarrow J(a,b) \times J(a',b') \tag{2.3}$$

by induction on $(a+b-a'-b')$.

If $(a+b-a'-b') = 0$, then we temporarily assume the surjectivity of the above map. We shall prove this in section 2.4.

If $(a+b-a'-b') > 0$, suppose that $b > b'$ (the case where $a > a'$ is similar). Denote by C'^+ the standard module dominating $J(a, b-1/n) \times J(a',b')$. We have

$$C^+ \cong [b] \times C'^+ \twoheadrightarrow [b] \times J(a, b-1/n) \times J(a',b') \twoheadrightarrow J(a,b) \times J(a',b').$$

The first surjection is by the induction hypothesis and the second one by the uniqueness of the Langlands quotient.

So the first part of (1) in the lemma is proved. Passing to the contragredients, we get another part of (1).

(2) Suppose $(a+b)/2 \geq (a'+b')/2$. We have the following commutative diagram

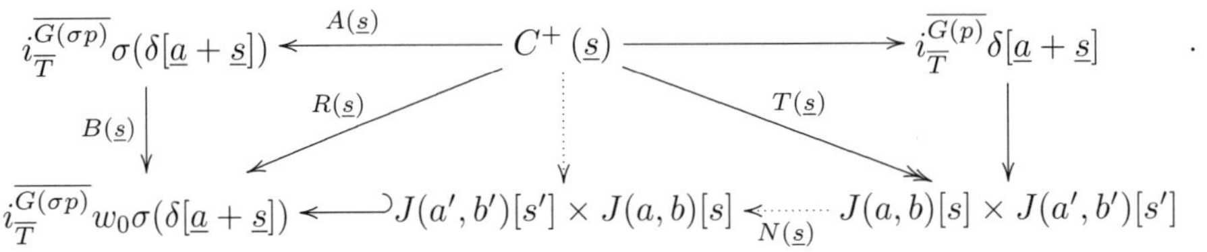

All operators are normalized intertwining operators. Their definitions should be clear from the diagram.

The operator $R(\underline{s}) = A(\underline{s})\, B(\underline{s})$ has (nonzero) image in $J(a',b') \times J(a,b)$ when $\underline{s} = \underline{0}$. And this image is the unique irreducible subrepresentation of $J(\delta',a',b') \times J(\delta,a,b)$. The surjectivity has just been proved for $T(\underline{s})$ at $\underline{s} = \underline{0}$. So $N(\underline{s})$ is defined at $\underline{s} = \underline{0}$.

Hence if $J(a,b) \times J(a',b')$ is irreducible, then $N(\underline{s})$ is an isomorphism at $\underline{s} = \underline{0}$. And the its inverse $N(\sigma, \sigma\pi, \sigma\underline{s})$ is holomorphic at $\underline{s} = \underline{0}$.

Conversely if both $N(\underline{s})$ and $N(\sigma, \sigma\pi, \sigma\underline{s})$ are defined at $\underline{s} = \underline{0}$, then they are isomorphisms. So $L(\delta, p, \underline{a})$ is both an irreducible subrepresentation and a quotient of $J(\delta,a,b) \times J(\delta',a',b')$ by part (1) of the lemma, which implies that C^+ has two isomorphic subquotients. By Proposition 1.19, $J(\delta,a,b)[\underline{s}] \times J(\delta',a',b')$ must be irreducible.

Passing to the contragredients, we see (2) is also true if $(a+b)/2 < (a'+b')/2$. □

2.2. Speh Modules (p-adic Case)

Assume F is p-adic. We keep the notation from the last section. In particular, σ stands for the simple reflection, and δ is a unitary irreducibly representation of $\overline{T}$.

LEMMA 2.3. (refer to [**MW89**, p. 618])
Suppose $a \geq 0$. *Put* $\pi = J(-a,0) \times 0$, *then*
(1) $N(\sigma, \pi, \underline{s})$ *is holomorphic at* $\underline{s} = \underline{0}$.
(2) The induced representation $I(\pi) = J(-a,0) \times [0]$ *is irreducible.*

PROOF. Applying lemma 2.1 to $\Pi = \sigma\pi$, we see (1) and (2) are equivalent hence we need only prove $N(\sigma, \pi, \underline{s})$ is holomorphic at $\underline{s} = \underline{0}$ by induction on a.

Assume δ is invariant under $\mathfrak{S}_r$. Otherwise, the lemma follows from Lemma 2.2 (3).

If $a = 0$, part (2) follows from Proposition 1.21 and hence (1) is also true.

Now suppose the lemma holds for $a \in \frac{1}{n}\mathbf{Z}$. We want to show it is also true for $a + 1/n$. Let $\pi' = J(-a-1/n, 0) \times 0$. First we have the following commutative diagram

$$\begin{array}{ccc}
J(-a-1/n,0)[s]\times[s'] & \xrightarrow{N(\sigma,\pi',\underline{s})} & [s']\times J(-a-1/n,0)[s] \quad . \\
\uparrow{\scriptstyle N((12))} & & \downarrow \\
J(-a,0)[s]\times[s-a-1/n]\times[s'] & \xrightarrow{A(\underline{s})=N((13))} & [s']\times[s-a-1/n]\times J(-a,0)[s] \\
\downarrow{\scriptstyle B(\underline{s})=N((12))} & & {\scriptstyle D(\underline{s})=N((12))}\uparrow \\
[s-a-1/n]\times J(-a,0)[s]\times[s'] & \xrightarrow{C(\underline{s})=N((23))} & [s-a-1/n]\times[s']\times J(-a,0)[s]
\end{array}$$

To show $N(\sigma, \pi', \underline{s})$ is holomorphic at $\underline{s} = \underline{0}$, we only need to discuss $D(\underline{s})$ since all other operators are holomorphic at $\underline{s} = \underline{0}$. (The holomorphy of $C(\underline{s})$ is by the induction hypothesis.)

If $a \neq 0$, then $D(\underline{s})$ is obviously holomorphic and the lemma is proved in this case.

If $a = 0$, we are essentially working on $\overline{GL(3)}$. Let $\underline{s} = (s, s')$. The operator

$$D(\underline{s}) = N((12)) : [s - 1/n] \times [s'] \times [s] \longrightarrow [s'] \times [s - 1/n] \times [s]$$

has at most a simple pole at $\underline{s} = \underline{0}$. Let $s_0 = s - s'$. Then

$$N_0 = \lim_{s_0 \to 0} s_0 N(\sigma, [-1/n] \times [0], \underline{s})$$

is defined with image denoted by $d \neq 0$. Define a partition $p = (2, 1)$. The module d is a representation of $\overline{G(p)}$. It must be the unique irreducible submodule of $[0] \times [-1/n]$. Otherwise, since the length of $[0] \times [-1/n]$ is two, N_0 would be surjective and hence an isomorphism which is impossible.

So $A(\underline{s})$ has at most one simple pole and the same is true for $N(\sigma, \pi', \underline{s})$. Define

$$A = \lim_{s_0 \to 0} s_0 A(\underline{s}) \ ; \qquad N = \lim_{s_0 \to 0} s_0 N(\sigma, \pi', \underline{s}) \ .$$

We just observed the image of $\lim_{s_0 \to 0} s_0 D(\underline{s_0})$ is $i_{\overline{G(p)}}^{\overline{G}} d$. We shall show

$$\operatorname{Hom}_{\overline{G}} \left(J(-1/n, 0) \times [0], i_{\overline{G(p)}}^{\overline{G}} d \right) = 0 \ . \tag{2.4}$$

So $N = 0$ and $N(\sigma, \pi', \underline{s})$ must be holomorphic at $\underline{s} = \underline{0}$. The lemma then follows.

We now prove (2.4). First recall that by definition,

$$J(-1/n, 0) \times [0] = i_{\overline{G(p)}}^{\overline{G}} L(p, (-1/n, 0, 0)).$$

If (ρ, V) is an admissible representation of $\overline{G}$ and M is a standard Levi subgroup of G, recall the Jacquet functor $r_{\overline{M}}^{\overline{G}} \rho$ as follows. Let $r_{\overline{M}}^{\overline{G}} V$ be the quotient space of V by the subspace generated by the vectors $\{\rho(n)v - v : v \in V, n \in N\}$. The group $\overline{M}$ acts it by

$$\left(r_{\overline{M}}^{\overline{G}} \rho \right)(m)(v) = \Delta_{M \cap (TN)}(m)^{-1/2} \rho(m)(v).$$

This is admissible representation of $\overline{M}$.

By the Frobenius reciprocity, we need to show

$$\operatorname{Hom}_{\overline{G(p)}} \left(r_{\overline{G(p)}}^{\overline{G}} i_{\overline{G(p)}}^{\overline{G}} L(p, (-1/n, 0, 0), d \right) = 0. \tag{2.5}$$

Since $\overline{G(p)}\backslash\overline{G}/\overline{G(p)} = \{1,(13),(123)\}$, by [**BZ77**, 5.2 theorem], the Jacquet module

$$r^{\overline{G}}_{\overline{G(p)}} i^{\overline{G}}_{\overline{G(p)}} L(p,(-1/n,0,0)$$

has three components:

$$L(p,(-1/n,0,0)); \tag{2.6}$$

$$i^{\overline{G(p)}}_{\overline{T}}(0,-1/n,0); \tag{2.7}$$

$$i^{\overline{G(p)}}_{\overline{T}}(0,0,-1/n). \tag{2.8}$$

By the multiplicity one of the Langlands quotient, $\mathrm{Hom}(L(p,(-1/n,0,0)),d) = 0$. Similarly, we see that there are no nontrivial homomorphisms from (2.7) or (2.8) to d. Now (2.5) follows. □

We now generalize the above lemma. Let a,b,a',b' be real numbers such that $a \le b$, $a' \le b'$ and $na, nb, na', nb' \in \mathbf{Z}$. Let $p = (n(b-a)+1, n(b'-a')+1)$ be a partition of r. Again, let δ be an irreducible unitary representation of $\overline{T}$ and σ be the simple transposition.

LEMMA 2.4. (refer to [**MW89**, p. 622])
Assume $w(\delta) = \delta$ for any $w \in \mathfrak{S}(p)$. Let $\pi = L(\delta, p, ([a,b],[a',b']))$.

1) If $b \ge b'$ or $a \ge a'$, then $N(\sigma,\pi,\underline{s})$ is holomorphic at $\underline{s} = \underline{0}$.

2) If (a,b) and (a',b') are not linked, or $\tau\delta \neq \delta$ for some $\tau \in \mathfrak{S}_r$, then $J(a,b) \times J(a',b')$ is irreducible and $N(\sigma,\pi,\underline{s})$ and $N(\sigma,\sigma\pi,\sigma\underline{s})$ are holomorphic at $\underline{s} = \underline{0}$.

PROOF. We have the following commutative diagram

$$\begin{array}{ccc} J(a,b)[s] \times J(a',b')[s'] & \xrightarrow{N(\sigma,\pi,\underline{s})} & J(a',b')[s'] \times J(a,b)[s] \\ \big\downarrow & & \big\downarrow \\ \begin{array}{c}[a+s] \times \cdots \times [b+s] \times \\ [a'+s'] \times \cdots \times [b'+s']\end{array} & \longrightarrow & \begin{array}{c}[a'+s'] \times \cdots \times [b'+s'] \\ \times [a+s] \times \cdots \times [b+s]\end{array} \end{array}$$

The arrow at the bottom is the product of $N((12), \delta[(j,j')], \underline{s})$ for $j \in [a,b]$, $j' \in [a',b']$. If each of them does not have a pole at $\underline{s} = \underline{0}$, then $N(\sigma,\pi,\underline{s})$ does not have a pole at $\underline{s} = \underline{0}$. By Corollary 1.13, $\underline{0}$ is a pole for some pair (j,j') if and only if

$$w(\delta) = \delta, \forall w \in \mathfrak{S}_r; \quad a - a' \in \frac{1}{n}\mathbf{Z}; \quad a < b', \quad a' \le b+1. \tag{2.9}$$

If the first two conditions are not satisfied, then $N(\sigma,\pi,\underline{s})$ is holomorphic at $\underline{s} = \underline{0}$, and similarly for $N(\sigma,\sigma\pi,\sigma\underline{s})$. The irreducibility follows from Lemma 2.1.

So we assume the first two conditions are satisfied in the proof. Assuming $b > b'$, we prove that $N(\sigma,\pi,\underline{s})$ is holomorphic at $\underline{0}$ by induction on $b'-a'$.

First suppose $b' = a'$. We may further suppose $a < b'$ (otherwise (2.9) is not satisfied). We have the injection

$$J(a,b) \hookrightarrow J(a,b') \times J\left(b' + \frac{1}{n}, b\right).$$

Consider the following commutative diagram (when $\underline{s}$ is around $\underline{0}$)

$$\begin{array}{ccc}
\begin{matrix}J(a,b)[s]\\ \times[b'+s']\end{matrix} & \xrightarrow{N(\sigma,\pi,\underline{s})} & \begin{matrix}[b'+s']\\ \times J(a,b)[s]\end{matrix}\\
\downarrow & & \downarrow\\
\begin{matrix}J(a,b')[s]\\ \times J(b'+1/n,b)[s]\\ \times[b'+s']\end{matrix} & \longrightarrow & \begin{matrix}[b'+s']\\ \times J(a,b')[s]\\ \times J(b'+1/n,b)[s]\end{matrix}\\
\searrow{\scriptstyle\text{holomorphic by lemma 2.1}} & & \nearrow{\scriptstyle\text{holomorphic by lemma 2.3}}\\
& \begin{matrix}J(a,b')[s]\\ \times[b'+s']\\ \times J(b'+1/n,b)[s]\end{matrix} &
\end{array}$$

So $N(\sigma,\pi,\underline{s})$ is holomorphic at $\underline{0}$ in this case.
Now suppose $a' < b'$, and consider the following commutative diagram

$$\begin{array}{ccc}
\begin{matrix}J(a,b)[s]\\ \times J(a',b')[s']\end{matrix} & \xrightarrow{N(\sigma,\pi,\underline{s})} & \begin{matrix}J(a',b')[s']\\ \times J(a,b)[s]\end{matrix}\\
\downarrow & & \downarrow\\
\begin{matrix}J(a,b)[s]\\ \times\delta[a'+s']\\ \times J(a'+1/n,b')[s']\end{matrix} & \longrightarrow & \begin{matrix}[a'+s']\\ \times J(a'+1/n,b')[s']\\ \times J(a,b)[s]\end{matrix}\\
\searrow{\scriptstyle\text{by the case just considered}} & & \nearrow{\scriptstyle\text{by induction hypothesis}}\\
& \begin{matrix}[a'+s']\\ \times J(a,b)[s]\\ \times J(a'+1/n,b')[s']\end{matrix} &
\end{array}$$

So $N(\sigma,\pi,\underline{s})$ is holomorphic at $\underline{0}$ when $b \geq b'$.

If $a \geq a'$, we get the same result by considering the contragredients. This finishes part (1) of the lemma.

For part (2), suppose (a,b) and (a',b') are not linked and show the holomorphy of $N(\sigma,\pi,\underline{s})$. We have observed at the beginning of the proof that we may assume (2.9) is satisfied. Then we see that either $a' \leq a$ or $b' \leq b$. The holomorphy of $N(\sigma,\pi,\underline{s})$ at $\underline{0}$ then follows from part one. By symmetry, $N(\sigma,\sigma\pi,\sigma\underline{s})$ is also holomorphic at $\underline{0}$. □

2.3. The Principal Lemma

Assume F is p-adic. For $i = 1,\dots,m$, we use the following notation:

$\Delta_i = [a_i, b_i]$ for $a_i, b_i \in \frac{1}{n}\mathbf{Z}$;
$p_i = n(b_i - a_i) + 1$;
$p = (p_1, \cdots, p_m)$;
$p'_1 = 0$; $\quad p'_i = \sum_{k=1}^{i-1} p_k$;
$r = \sum_{i=1}^{m} p_i$;
$\underline{s} = (s_1, \cdots, s_m) \in \mathbf{C}^m$;
$\lambda(\underline{s}) = (b_1 + s_1, b_1 + s_1 - 1/n, \cdots, a_1 + s_1, b_2 + s_2, \cdots, a_m + s_m)$;
$w \in \mathfrak{S}_r$ reverses the order on each interval $[p'_i + 1, p'_{i+1}]$: $w(p'_i + j) = p'_{i+1} + 1 - j$.

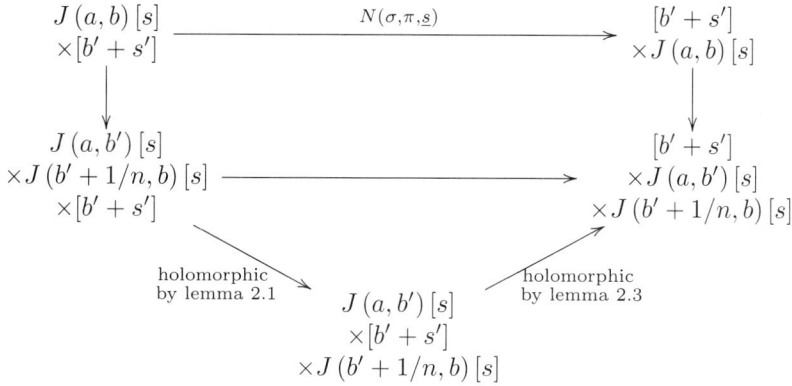

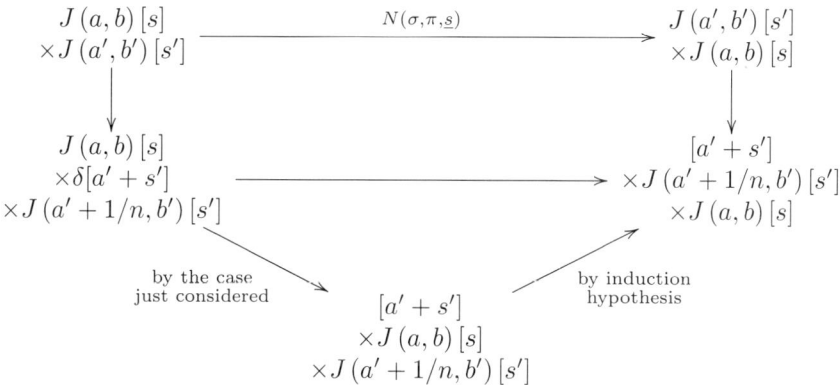

We say that the segment $[a,b]$ dominates $[a',b']$, and write $[a,b] \geq [a',b']$, if we have either $b > b'$ or $b = b'$ and $a \geq a'$.

Fix an irreducible unitary representation δ of $\overline{T}$ such that $\sigma\delta = \delta$ for any $\sigma \in \mathfrak{S}(p)$. Let $J_i = J(a_i, b_i)$. Remark by our convention, J_i is an irreducible representation of $\overline{G_i}$, where G_i is the standard Levi corresponding to the partition

$$r = 1 \times p_i' + p_i + 1 \times (r - p_{i+1}') = \underbrace{1 + \cdots + 1}_{p_i'} + p_i + \underbrace{1 + \cdots + 1}_{r - p_{i+1}'}.$$

Define

$$J = J_1 \otimes J_2 \otimes \cdots \otimes J_m = L(\delta, p, \lambda(0))$$

which is an irreducible representation of $\overline{G(p)}$. Remark this is not a tensor product of representation. But by Lemma 1.24, we still have for any $\sigma \in \mathfrak{S}_m$

$$\sigma(J_1 \otimes \cdots \otimes J_m) = J_{\sigma^{-1}(1)} \otimes \cdots \otimes J_{\sigma^{-1}(m)}.$$

The operator $N(w, \delta, \lambda(\underline{s})) : I(\delta, \lambda(\underline{s})) \longrightarrow I(\delta, w\lambda(\underline{s}))$ is defined with image $I(J, \underline{s})$.

Fix $w' \in \mathfrak{S}_r$ such that if $\lambda'(\underline{s}) = w'^{-1}\lambda(\underline{s})$, we have

$$\mathrm{Re}\left(\lambda'(\underline{0})_i - \lambda'(\underline{0})_j\right) \geq 0 \quad \text{for all } 1 \leq i \leq j \leq r.$$

Put $\delta' = w'^{-1}\delta$. Then $N(ww', \delta', \lambda'(\underline{s}))$ is defined as a meromorphic operator. It factors through $N(w, \delta, \lambda(\underline{s}))$. So its image is included in $I(J, \underline{s})$.

Introduce the following condition:

$$\begin{array}{l} \text{For any } 1 \leq i < j \leq m, \text{ we have } \tau\delta \neq \delta \text{ for } \tau = (p_{i+1}', p_{j+1}'), \\ \text{or } (a_i, b_i) \text{ is not linked with } (a_j, b_j), \text{ or } [a_i, b_i] \text{ dominates } [a_j, b_j]. \end{array} \tag{2.10}$$

LEMMA 2.5. (refer to [**MW89**, p. 633])

Suppose $\sigma \in \mathfrak{S}_r$ *and* σ *is increasing on each interval* $[p_i' + 1, p_{i+1}']$.

1) There is a meromorphic operator, denoted by $N_\sigma(\underline{s})$, *such that the following diagram commutes*

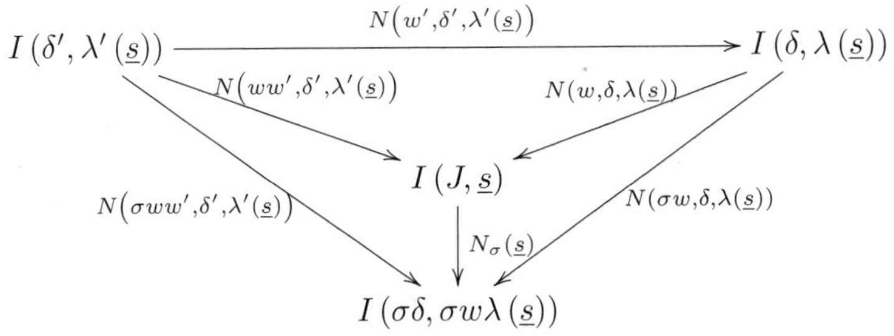

2) If condition (2.10) is verified, then all the above operators are holomorphic at $\underline{s} = \underline{0}$. *The operator* $N(ww', \delta', \lambda(\underline{0}))$ *has* $I(J, \underline{0})$ *as its image and* $N_\sigma(\underline{0}) \neq 0$.

PROOF. Define $N_\sigma(\underline{s}) = N(\sigma, \delta, w\lambda(\underline{s}))\,|_{I(J,\underline{s})}$. Then (1) follows.

By Proposition 1.17, all operators starting from $I(\delta', \lambda'(\underline{s}))$ are holomorphic at $\underline{0}$, and none of them is zero.

Suppose we can show that

$$N(ww', \delta', \lambda'(\underline{0})) \text{ is surjective.} \tag{2.11}$$

Since all the spaces of the above induced representations can be realized as an induced space independent of $\underline{s}$, there is a right inverse (as a linear operator) $\iota(\underline{s})$:

$I(J,\underline{s}) \longrightarrow I(\delta',\lambda'(\underline{s}))$, analytic at $\underline{0}$. Thus $N_\sigma(\underline{s}) = N(\sigma ww',\delta',\lambda'(\underline{s})) \circ \iota(\underline{s})$ is analytic at $\underline{0}$. Commutativity of the diagram implies the other assertions.

So the only thing left is to show the surjectivity of $N(ww',\delta',\lambda'(\underline{0}))$.

Suppose $\tau \in \mathfrak{S}_m$ such that $\Delta_{\tau(1)} \geq \Delta_{\tau(2)} \geq \cdots \geq \Delta_{\tau(m)}$. Consider

$$N\left(\tau, \tau^{-1}J, \tau^{-1}\underline{s}\right) : I\left(\tau^{-1}J, \tau^{-1}\underline{s}\right) \longrightarrow I(J,\underline{s}).$$

Suppose $(i',j') \in \mathrm{inv}(\tau)$, put $i = \tau i', j = \tau j'$. We have $i > j$ and $\Delta_i \geq \Delta_j$. Condition (2.10) implies that either (a_i, b_i) and (a_j, b_j) are not linked or $\sigma\delta \neq \delta$ for $\sigma = (p'_{i+1}, p'_{i+1} + 1)$ and hence the above operator is holomorphic and is an isomorphism at $\underline{0}$ by lemma 2.4.

It remains to show that $N\left(\tau^{-1}ww',\delta',\lambda'(\underline{s})\right) = N(w_\tau w'_\tau, \delta', \lambda'(\underline{s}))$ is surjective by the following diagram. Here $w_\tau = \tau^{-1}w\tau, w'_\tau = \tau^{-1}w'$.

$$
\begin{array}{ccc}
I(\delta',\lambda'(\underline{s})) & & \\
\Big\downarrow{\scriptstyle N\left(\tau^{-1}ww',\delta',\lambda'(\underline{s})\right)} & \!\!\!\!\!\!\!\!\!\!\!\!\!\!\!\searrow{\scriptstyle N\left(ww',\delta',\lambda'(\underline{s})\right)} & \\
I\left(\tau^{-1}J,\tau^{-1}\underline{s}\right) & \xrightarrow[\;\; N\left(\tau,\tau^{-1}J,\tau^{-1}\underline{s}\right)\;\;]{} & I(J,\underline{s})
\end{array}
$$

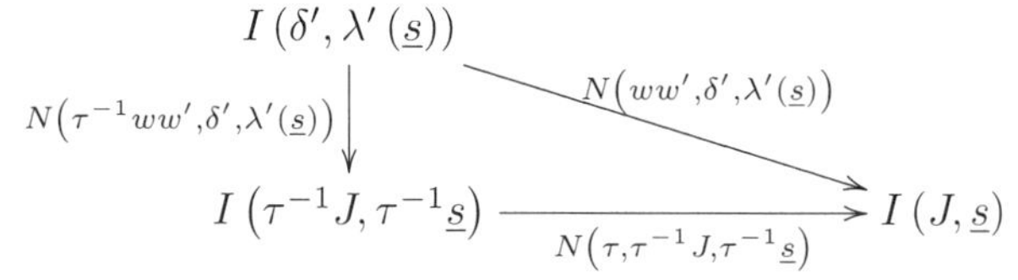

So in the remaining part of the proof, we assume that $\Delta_1 \geq \Delta_2 \geq \cdots \geq \Delta_m$.

We are going to show (2.11) by induction on m. Suppose $u',v',u,v \in \mathfrak{S}_r$ such that

- $u' = w'$ on $w'^{-1}(\{1,\cdots,p_1\})$;
- u' is increasing on $\{1,\cdots,r\} - w'^{-1}(\{1,\cdots,p_1\})$;
- $v' = w'u'^{-1}$;
- $u = w$ on $\{1,\cdots,p_1\}$;
- $u =$ identity on $\{p_1+1,\cdots,r\}$;
- $v = wu^{-1}$.

Then we have

$$ww' = vv'uu';$$
$$\tau(uu'\delta') = uu'\delta', \quad \forall \tau = (ij), 1 \leq i < j \leq p_1;$$
$$N(ww',\delta',\lambda'(\underline{s})) = N(vv',uu'\lambda'(\underline{s})) \circ N(uu',\delta',\lambda'(\underline{s})).$$

We can write
$$uu'\lambda'(\underline{s}) = (a_1 + s_1, \cdots, b_1 + s_1, \lambda'^v(\underline{s}^v))$$
where $\underline{s}^v = (s_2, s_3, \cdots, s_m)$. And $\lambda'^v(\underline{s}^v)$ is the analogue of $\lambda'(\underline{s})$ relative to the segments $\Delta_2, \cdots, \Delta_m$.

Suppose we can demonstrate that

(2.12)   The image of $N(uu',\delta',\lambda'(\underline{0}))$ is $J_1 \times_{uu'\delta'} I(\lambda'^v(\underline{0}))$.

The lemma then follows from (2.12) and the induction hypothesis

$$N(vv',uu'\delta',uu'\lambda'(\underline{0})) \cdot (J_1 \times_{uu'\delta'} I(\lambda'^v(\underline{0}))) = J_1 \times_{ww'\delta'} (J_2 \times \cdots \times J_m).$$

Now we prove (2.12). For $i = 1, \ldots, p_1$, put $j_i = w'^{-1}(i)$, we have $\lambda'_{j_i} = b_1 + (1-i)/n$. Put $k_i = \sup\{z : \lambda'(\underline{0})_z = b_1 + (1-i)/n\}$ and formally $k_0 = 0$. Then we have

$$0 = k_0 < j_1 \leq k_1 < j_2 \leq k_2 < \cdots < j_{p_1} \leq k_{p_1} < \cdots .$$

Denote by u^i the element of $\mathfrak{S}_r$ such that

$u^i(j_{p_1}) = k_{i-1}+1,\ u^i(j_{p_1-1}) = k_{i-1}+2,\ \cdots,\ u^i(j_i) = k_{i-1}+p_1+1-i,$

u^i is increasing on $\{1,\ldots,r\}-\{j_i,j_{i+1},\cdots,j_{p_1}\}$.

Put

$$\mu^i(\underline{s}) = \left(\lambda'(\underline{s})_1,\cdots,\lambda'(\underline{s})_{k_{i-1}}\right);\quad \nu^i(\underline{s}) = (\cdots,\lambda'(\underline{s})_z,\cdots),$$

where z runs over the series $\left(u^i\right)^{-1}(k_{i-1}+p_1+2-i),\cdots,\left(u^i\right)^{-1}(r)$. Observe that

$$\sigma u^i\delta' = u^i\delta' \quad \forall\sigma = (j,j+1), k_{i-1}+1\le j< k_{i-1}+p_1-i.$$

We prove by downward induction on i that we have

(2.13) The image of $N\left(u^i,\delta',\lambda'(\underline{0})\right)$ is $\mu^i(\underline{0})\times_{u^i\delta'} J\left(a_1,b_1+\dfrac{1-i}{n}\right)\times\nu^i(\underline{0})$.

We see when $i=1$, (2.13) becomes (2.12).

Put formally $u^i=\text{id}$ for $i=p_1+1$. If $i<p_1$ and suppose (2.13) is proved for $i+1$. Put $\tau^i = u^i\left(u^{i+1}\right)^{-1}$, and observe that

$N\left(u^i,\delta',\lambda'(\underline{s})\right) = N\left(\tau^i,u^{i+1}\delta',u^{i+1}\lambda'(\underline{s})\right)\circ N\left(u^{i+1},\delta',\lambda'(\underline{s})\right);$

$\tau^i(j_i) = k_{i-1}+p_1+1-i;$

$\tau^i(k_i+1) = k_{i-1}+1, \tau^i(k_i+2) = k_{i-1}+2,\cdots,\tau^i(k_i+p_1-i) = k_{i-1}+p_1-i;$

τ^i is increasing on other elements.

The pairs inverted by τ^i are

$$\begin{array}{lll}(z,j_i) & \text{for} & k_{i-1}+1\le z<j_i,\\ (z,z') & \text{for} & k_{i-1}+1\le z<k_i+1\le z'\le k_i+p_1-i\ .\end{array}$$

We also have

$$u^{i+1}\lambda'(\underline{s}) = \left(\mu^{i+1}(\underline{s}),a_1+s_1,\cdots,b_1+s_1-\tfrac{i}{n},\nu^{i+1}(\underline{s})\right);$$
$$u^{i+1}\lambda'(\underline{0})_z \ge u^{i+1}\lambda'(\underline{0})_{z'} \text{ for } (z,z')\in\text{inv}\left(\tau^i\right).$$

So $N\left(\tau^i,u^{i+1}\delta',u^{i+1}\lambda'(\underline{s})\right)$ is holomorphic at $\underline{s}=\underline{0}$, hence

(2.14)

$$\text{Image}N\left(u^i,\delta',\lambda'(\underline{0})\right) = N\left(\tau^i,u^{i+1}\delta',u^{i+1}\lambda'(\underline{0})\right)\text{Image}N\left(u^{i+1},\delta',\lambda'(\underline{0})\right).$$

The second image is understood by the induction hypothesis if $i<p_1$. By convention u^{p_1+1} =id. So if $i=p_1$, the operator $N(u^{i+1},\delta',\lambda'(\underline{0}))$ is the identity.

We need to study $N\left(\tau^i,u^{i+1}\delta',u^{i+1}\lambda'(\underline{0})\right)$ on the second image.

If $i<p_1$, it is the composition of the following operators (For simplicity, denote $\lambda'_i=\lambda'(0)_i$. And $\boxed{\lambda'_{j_i}}$ indicates the term inside should be omitted.)

$$\begin{aligned}
&\mu^{i+1}(0)\times J(a_1,b_1-i/n)\times\nu^{i+1}(0)\\
=\ &\mu^i(0)\times\lambda'_{k_{i-1}+1}\times\cdots\times\lambda'_{k_i}\times J(a_1,b_1-i/n)\times\nu^{i+1}(0)\\
\xrightarrow{N_1}\ &\mu^i(0)\times\lambda'_{k_{i-1}+1}\times\cdots\times\boxed{\lambda'_{j_i}}\times\cdots\times\lambda'_{k_i}\times\lambda'_j\times J(a_1,b_1-i/n)\times\nu^{i+1}(0)\\
\xrightarrow{N_2}\ &\mu^i(0)\times\lambda'_{k_{i-1}+1}\times\cdots\times\boxed{\lambda'_{j_i}}\times\cdots\times\lambda'_{k_i}\times J(a_1,b_1+(1-i)/n)\times\nu^{i+1}(0)\\
\xrightarrow{N_3}\ &\mu^i(0)\times J(a_1,b_1+(1-i)/n)\times\lambda'_{k_{i-1}+1}\times\cdots\times\boxed{\lambda'_{j_i}}\times\cdots\times\lambda'_{k_i}\times\nu^{i+1}(0)\\
=\ &\mu^i(0)\times J(a_1,b_1+(1-i)/n)\times\nu^i(0).
\end{aligned}$$

In the above diagram, N_1 is an isomorphism since $\lambda'_j = \lambda'_l$ for $j_i < l \leq k_i$. By the definition of k_i and the condition $\Delta_1 \geq \Delta_2 \geq \cdots \geq \Delta_m$, $\left(a_1, b_1 + \frac{1-i}{n}\right)$ and (z_h) are not linked for $j_i + 1 \leq h \leq k_i$, hence N_3 is an isomorphism by lemma 2.4. N_2 is in fact the map of the Langlands quotient. Hence (2.13) is proved for $i < p_1$ (under the induction hypothesis).

If $i = p_1$, $J\left(\delta_1, a_1, b_1 - \frac{i}{n}\right)$ disappears. And $\tau^{p_1} = u^{p_1}$ which sends j_{p_1} to $k_{p_1 - 1} + 1$ and is increasing on other numbers. Also observe in this case that $0 < \lambda'_{k_{p_1}+1} - \lambda'_{j_{p_1}} < 1$ and $\nu^{p_1+1}(0)_1 = \lambda'_{j_{p_1}} = a_1$. So we have the surjection

$$N(\tau^{p_1}, \delta', \lambda') : \lambda' \twoheadrightarrow \mu^{p_1}(0) \times a_1 \times \nu^{p_1}(0),$$

which proves (2.13) in this case. □

Let $(ij) \in \mathfrak{S}_m$ permuting intervals corresponding to the partition p. The proof of the corollary is similar to that in nonmetaplectic case.

COROLLARY 2.6. (refer to [**MW89**, p. 638])
Suppose for any $i, j \leq m$, either the segments (a_i, b_i) and (a_j, b_j) are not linked or $(ij)\delta \neq \delta$. Then $J(a_1, b_1) \times \cdots \times J(a_m, b_m)$ is irreducible. □

2.4. Generalization and Completion of the Proof

Assume F is a local field (p-adic or complex). We keep the notation on page 27 in the last section. The definition of a segment $[a, b]$ for the complex case is the same as that for p-adic case (see (2.1)). The following lemma follows from Proposition 1.17.

LEMMA 2.7. (refer to [**MW89**, p. 639])
Suppose δ is an irreducible unitary representation of $\overline{T}$ and let $\underline{s} \in \mathbf{C}^r, \sigma \in \mathfrak{S}_r$. Suppose for all $1 \leq i < j \leq r$, $Re(s_i - s_j) > -1/n$. Then $N(\sigma, \delta, \cdot)$ is holomorphic at $\underline{s}$ and not zero. □

The main result in this chapter is the following proposition.

PROPOSITION 2.8. (refer to [**MW89**, p. 640])
Suppose $\sigma \in \mathfrak{S}_r, \underline{s} \in \mathbf{C}^m, w' \in \mathfrak{S}_r$. Define $\lambda(\underline{s})$ as in the previous section. Put $\delta' := w'^{-1}(\delta)$, $\lambda'(\underline{s}) := w'^{-1}(\lambda(\underline{s}))$. Suppose the following conditions are satisfied:
(a) for $1 \leq i < j \leq r$, $\mathrm{Re}\left(\lambda'(\underline{s})_i - \lambda'(\underline{s})_j\right) > -1/n$;
(b) σ is increasing on each interval $[p'_i + 1, p'_{i+1}]$;
(c) For all $1 \leq i \leq j \leq m$, $\mathrm{Re}(s_i - s_j) > -1/n$.
Then there is a meromorphic operator $N_\sigma(\underline{t})$ $(\underline{t} \in \mathbf{C}^m)$, such that the following diagram commutes.

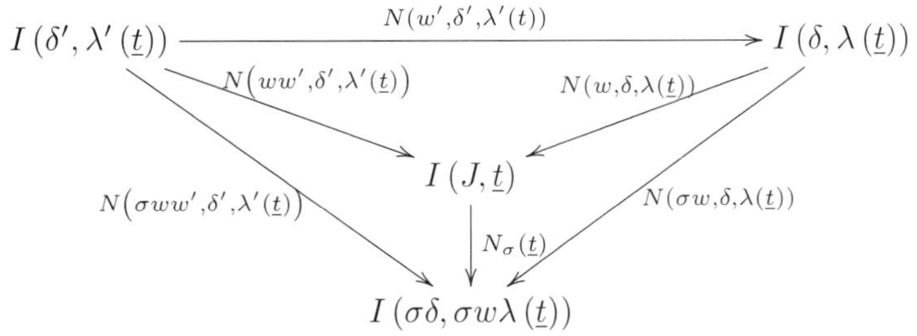

Furthermore, all the above operators are holomorphic at $\underline{s}$. Finally, the operator $N(ww', \delta', \lambda'(\underline{s}))$ is surjective and $N_\sigma(\underline{s}) \neq 0$.

PROOF. We prove for p-adic case first and leave the complex case to the next section. The proof is easier in p-adic case since we only consider representations induced from the ones of the diagonal subgroup.

Let w'' be an element in $\mathfrak{S}_r$ such that if $\lambda''(\underline{t}) = w''^{-1}w'^{-1}\lambda(\underline{t})$ then $\mathrm{Re}(\mu''(\underline{t})_i - \mu''(\underline{t})_j) \geq 0$ for $1 \leq i \leq j \leq r$. Put $\delta'' = w''^{-1}w'^{-1}\delta$. By lemma 2.7, we get isomorphisms

$$I(\delta', \lambda'(t)) \underset{N(w'',\pi'',\lambda''t)}{\overset{N(w''^{-1},\pi',\lambda'(t))}{\rightleftarrows}} I(\delta'', \lambda''(t)) \,. \tag{2.15}$$

Condition (c) gives the condition (2.10). Applying Lemma 2.5, we get the assertions in the proposition with w' replaced by $w'w''$. Now the proposition follows from (2.15). □

Let F be p-adic. Now we fill the gap as promised in the proof of lemma 2.2. If $\sigma\delta = \delta$ for any $\sigma \in \mathfrak{S}_r$, we need to show the surjectivity (2.3) when $a+b = a'+b'$ and $n(b'-b) \in \mathbf{Z}$. We accomplish this by showing that $J(a,b) \times J(a',b')$ is irreducible. After a twist of the character $|\det|^{-\frac{a+b}{2}}$, we many assume $a+b = a'+b' = 0$. We start with an easy lemma.

LEMMA 2.9. *Suppose $a, b \in \frac{1}{n}\mathbf{Z}$ and $0 \leq a \leq b$. Let δ be an irreducible representation of $\overline{T}$ fixed by $\mathfrak{S}_r$. Then the image of $J(a+1/n, b) \times J(0,a)$ under the map $N((12))$ is $J(0,b)$ where (12) transposes the blocks associated with the partition $(na+1, n(b-a)+1)$.*

PROOF. Denote by w_1, w_2 and w the permutations reversing the orders of segments $[0,a]$ $[a+1/n, b]$ and $[0,b]$ respectively. Then

$$\begin{aligned} &N((12))\,(J(a+1/n,b) \times J(0,a)) \\ &\quad= N((12))\,N(w_1w_2)\,([b] \times \cdots \times [0]) \\ &\quad= N(w)\,([b] \times \cdots \times [0]) \\ &\quad= J(0,b)\,. \end{aligned}$$

The lemma follows. □

LEMMA 2.10. *Let δ be an irreducible representation of $\overline{T}$ fixed by $\mathfrak{S}_r$. Assume $a, b \in \frac{1}{2n}\mathbf{Z}$. Then $J(a,-a) \times J(b,-b)$ is irreducible.*

PROOF. Suppose $G = GL(r)$, $r = 2an + 2bn + 2$. Denote by $w' \in \mathfrak{S}_r$ the element for which $w'^{-1}(-a, -a+1/n, \cdots, a, -b, \cdots, b)$ is non-increasing and that if we denote $(x_1, \cdots, x_r) = (-a, -a+1/n, \cdots, a, -b, \cdots, b)$ then $w'(i) < w'(j)$ for $i < j$, $x_i = x_j$. Let

$$\lambda = w'^{-1}(-a, -a+1/n, \cdots, a, -b, \cdots, b)\,.$$

We prove the following statements by induction on r:

(1) This lemma.
(2) The map $N(w', \delta', \lambda) : I(\delta', \lambda) \longrightarrow J(\delta, -a, a) \times J(\delta, -b, b)$ is surjective.
(3) lemma 2.2.
(4) lemma 2.5 (the surjection (2.11)).

We first show that (2) implies (1) from which we deduce immediately that (1) and (2) are equivalent. Suppose we have the surjection as in statement (2). If L is the unique quotient of $I(\delta, \lambda)$, then it is also a quotient of $J(-a, a) \times J(-b, b)$ which must be unique. Applying statement (2) to the contragredient $\check{\delta}$ and segments $(-a, a)$ and $(-b, b)$ and using the duality, we see L is also a subrepresentation of $J(-a, a) \times J(-b, b)$. Since $J(-a, a) \times J(-b, b)$ is a quotient of $I(\delta', \lambda)$, L has multiplicity one in $J(-a, a) \times J(-b, b)$. So $J(-a, a) \times J(-b, b)$ must be irreducible and isomorphic to L.

Now we prove statements (2), (3) and (4) by induction. When $b = 0$, we have
$[a] \times [a - \frac{1}{n}] \times \cdots \times [\frac{1}{n}] \times [0] \times [0] \times [-\frac{1}{n}] \times \cdots \times [-a]$
$\to ([\frac{1}{n}] \times \cdots \times [a - \frac{1}{n}] \times [a]) \times [0] \times ([-a] \times \cdots \times [-\frac{1}{n}] \times [0])$.
The actual image is

$$J\left(\frac{1}{n}, a\right) \times [0] \times J(-a, 0) \cong J\left(\frac{1}{n}, a\right) \times J(-a, 0) \times [0] \to J(-a, 0) \times J\left(\frac{1}{n}, a\right) \times [0].$$

The above isomorphism follows from Lemma 2.3 for the non-archimedean case. Note that when $\sigma\delta \neq \delta$ for any $\sigma \in \mathfrak{S}_r$, the proof of Lemma 2.3 did not use the lemma we are proving. The image of the above arrow is $J(-a, a) \times [0]$ by lemma 2.9.

This proves (2) when $b = 0$. Similarly we can prove (2) when $a = 0$.

We observe that in the previous sections, we did everything within $G = GL(r)$. So by Lemma 2.5 and Lemma 2.2, we may assume all the four statements are true for r. Now we are proving (2) for $r + 1$, then (1), (3), (4) for $r + 1$ follow from (2) for $r + 1$.

If $a \leq b$, we have

$$\begin{aligned} & I(\delta, \lambda) && \text{(the last factor is } [-b]) \\ \twoheadrightarrow\ & J(-a, a) \times J(-b + \tfrac{1}{n}, b) \times [-b] && \text{(induction hypothesis (statement (4)))} \\ \overset{(23)}{\twoheadrightarrow}\ & J(-a, a) \times J(-b, b)\ . && \text{(lemma 2.9)} \end{aligned}$$

Similarly we can prove the surjectivity when $a > b$. □

2.5. The Complex Case

Suppose F is the field of complex numbers and prove Proposition 2.8. The covering group $\overline{G}$ is a trivial n-fold covering of $GL(r, \mathbf{C})$. If $n = 1$, Proposition 2.8 is [**MW89**, I.11 Proposition]. We prove the proposition by reducing to the case when $n = 1$.

Keep notation in Proposition 2.8 (which are defined on page 27). We also use the following notation (only in this section). Suppose $x, y, z, x_i, y_i \in \mathbf{R}$, $i \leq k$, such that $x - y, x_i - y_i \in \frac{1}{n}\mathbf{Z}$. Let

- $[x, y]$ = the sequence $(x, x + \frac{1}{n}, \cdots, y - \frac{1}{n}, y)$ if $x \leq y$;
- $[x, y]$ = the sequence $(x, x - \frac{1}{n}, \cdots, y + \frac{1}{n}, y)$ if $x > y$;
- $\langle x, y : z\rangle$ = the sequence $\{z + i : i \in \mathbf{Z}, z + i \text{ is between } x \text{ and } y\}$ in increasing order if $x \leq y$; decreasing if $x > y$.
- $\cup_{i=1}^{k}[x_i, y_i] = [x_1, y_1] \cup [x_2, y_2] \cup \cdots \cup [x_k, y_k]$ = the sequence formed by concatenation (the order matters).

For $i, k \in \mathbf{Z}$ such that $i \leq 0$ and $k > 0$, denote by $\vartheta(i, k)$ the number of positive integers $j \leq k$ such that $j \equiv i \mod n$. Then $\vartheta(i, k)$ is the largest integer no bigger than $(k - i)/n$. Observe that $\langle b_j, a_j : b_j - i/n\rangle$ contains $\vartheta(i, p_j)$ elements.

Recall for $i \leq m$, we have defined $p_i = n(b_i - a_i) + 1$ and $(p_1, \cdots, p_m)$ gives a partition of r. For each p_j, define the partition $q^j = (q_1^j, q_2^j, \cdots, q_{m(j)}^j)$ where $q_i^j = \vartheta(i, p_j)$. It follows that $m(j) = n$ if $q^j \geq n$ and $m(j) = q^j$ other wise. Notice the sequence $q = (q_i^j : j \leq m, i \leq m(j))$ forms a partition of r which is a sub-partition of p. Define

$$\lambda'(\underline{0}) = \left(\cup_{i=0}^{n-1} \langle b_1, a_1 : b_1 - i/n \rangle \right) \cup \cdots \cup \left(\cup_{i=0}^{n-1} \langle b_m, a_m : b_m - i/n \rangle \right) \in \mathbf{C}^r.$$

Note that $\lambda'(\underline{0})$ associated with the partition q is an analog of $\lambda(\underline{0})$ associated with the partition p. For $\underline{s} \in \mathbf{C}^m$, define $\lambda'(\underline{s})$ by $\lambda'(\underline{s})_i = \lambda'(\underline{0})_i + s_j$ if i is in the interval $[p_j' + 1, p_{j+1}']$.

Define a permutation τ in $\mathfrak{S}_r$ as follows. For $1 \leq l < m$, $0 \leq k < n$, and $u \in \mathbf{Z}$ such that $u \geq 0$ and $nu + k \leq p_i$, let

$$\tau \left(\sum_{i=1}^{l} (n(b_i - a_i) + 1) + nu + k \right) = \sum_{i=1}^{l} (n(b_i - a_i) + 1) + \sum_{i=0}^{k-1} \vartheta(i, p_{l+1}) + (u + 1).$$

By convention, $\sum_{i=0}^{k-1} = 0$ if $k = 0$. Observe that τ sends $\lambda(\underline{0})$ to $\lambda'(\underline{0})$.

If p is a partition of r, denote by $w(p)$ the element in $\mathfrak{S}_r$ inversing the order on each interval $[p_i' + 1, p_{i+1}']$. By our previous definition of w, we see $w = w(p)$. Let $w_0 = w(q)$. Denote by w_1 the permutation in $\mathfrak{S}_r$ given by

$$w_1(p_j' + q_1^j + \cdots + q_i^j + k) = (p_j' + q_{i+1}^j + \cdots + q_{m(j)}^j + k).$$

The permutation w_1 stabilize the partition p but inverses order of each sub-partition $(q_1^j, q_2^j, \cdots, q_{m(i)}^j)$. Observe that $w = w_0 w_1 = w_1 w_0$ and $\tau w = w \tau = w_0 w_1 \tau$.

Observe that in the commutative diagram

$$
\begin{array}{ccc}
I(\delta, \lambda(\underline{t})) & \xrightarrow{\ N(w_1 \tau)\ } & I(\delta, \lambda'(t)) \\
\downarrow{\scriptstyle N(w)} & & \downarrow{\scriptstyle N(w_0)} \\
I(\delta, w\lambda(\underline{t})) & \xrightarrow{\ N(\tau)\ } & I(\delta, w_0\lambda'(\underline{t}'))
\end{array}
\quad,
$$

both horizontal arrows are isomorphism. Let J' be the irreducible subrepresentation of $i_{\overline{T}}^{\overline{G(q)}} \delta[w_0 \lambda'(\underline{0})]$. The images of $N(w)$ and $N(w_0)$ are $I(J, \underline{t})$ and $I(J', \underline{t}')$ respectively. So they must be isomorphic. Consider the following commutative diagram

$$
\begin{array}{ccccc}
I(\delta', \lambda'(t)) & \xrightarrow{\ N(w')\ } & I(\delta, \lambda(\underline{t})) & \xrightarrow{\ N(w_1\tau)\ } & I(\delta, \lambda'(t)) \\
 & \searrow{\scriptstyle N(ww')} & \downarrow{\scriptstyle N(w)} & & \downarrow{\scriptstyle N(w_0)} \\
 & & I(J, \underline{t}) & \xrightarrow{\ N(\tau)\ } & I(J', \underline{t}') \\
 & \swarrow{\scriptstyle N(\sigma ww')} & \downarrow{\scriptstyle N(\sigma)} & \nearrow{\scriptstyle N(\sigma\tau^{-1})} & \\
 & & I(\sigma\delta, \sigma w\lambda(\underline{s})) & &
\end{array}
\quad.
$$

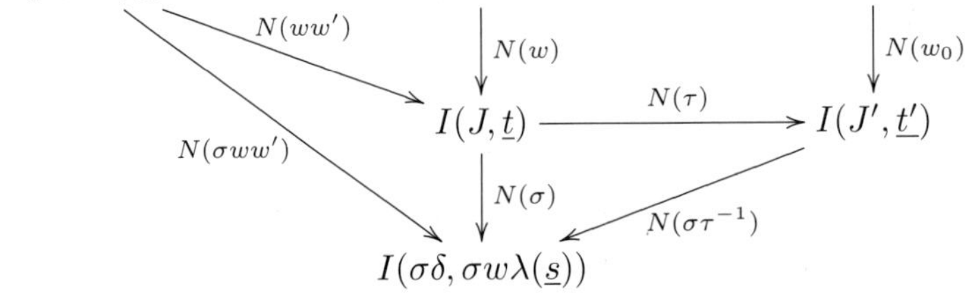

Observe that $\sigma\tau^{-1}$ is increasing on each interval $[p_j' + q_1^j + q_2^j + \cdots + q_i^j + 1, p_j' + q_1^j + q_2^j + \cdots + q_{i+1}^j]$. Applying [**MW89**, I.11 Proposition] to $(1,1)$, $(1,3)$, $(2,3)$ and $(3,2)$ entries in the above diagram, we see that $N_{\sigma\tau^{-1}} = N(\sigma\tau^{-1})|_{I(J,\underline{t}')}$ is surjective, and that the operators $N(w_1\tau)N(w')$, $N(\tau)N(ww')$, $N(\sigma ww')$ and $N(\sigma\tau^{-1})$ are all holomorphic at $\underline{t} = \underline{s}$. Since both $N(w_1\tau)$ and $N(\tau)$ are isomorphisms, Proposition 2.8 follows.

CHAPTER 3

Spectrum Associated with the Diagonal Subgroup

3.1. The Global Metaplectic Group

Let F be a number field and $\mathbf{A} = \mathbf{A}_F$ be the ring of adeles. Denote $G = GL(r)$. By a metaplectic cover $\overline{G} = \overline{G(\mathbf{A})}$ of $G(\mathbf{A})$, we mean a central extension $\overline{G(\mathbf{A})}$ of $G(\mathbf{A})$ by a finite abelian group μ whose restriction to $G(F)$ splits.

As in the local case, let T, H and N be the subgroup of diagonal matrices, diagonal matrices with determinant one and unipotent upper triangular matrices respectively.

Fix a metaplectic cover $\overline{G(\mathbf{A})}$. There is an integer n such that F contains all n-th roots of unity and such that the restriction of the cover to $SL(r, \mathbf{A})$ is given by a power of the global n-th Hilbert symbol.

Specifically, the group $\mu_n = [\overline{H(\mathbf{A})}, \overline{H(\mathbf{A})}]$ is isomorphic to the group μ_n of all n-th roots of unity. And μ_n is isomorphic to a subgroup $F^\times$. There is an integer k, relatively prime to n, satisfying the following properties. For each place v of F, Let $c_v(x, y) = (x, y)_v^k$, where $(\cdot, \cdot)_v$ is the local n-th Hilbert symbol on F_v. Let $\overline{G(F_v)}$ to be the cover of $G(F_v)$ as in Section 1.1. As seen in that section, there is a finite set V of places, including all archimedean ones, such that for every $v \notin V$, $\overline{G(F_v)}$ splits over $K_v = G(\mathfrak{O}_v)$. And the splitting isomorphism $\kappa_v : K_v \to \overline{G(F_v)}$ is uniquely defined. Let $K_v^* = \kappa(K_v)$. Denote by $\prod_v \overline{G(F_v)}[K_v^*]$ the restrictive direct product of $\overline{G(F_v)}$ relative to K_v^* $(v \notin V)$. Then we have the following isomorphism between topological groups:

$$\overline{G(\mathbf{A})} \cong \left(\prod_v \overline{G(F_v)}[K_v^*] \right) / \Xi, \tag{3.1}$$

where Ξ is the subgroup generated by the elements of the form

$$(\cdots, \xi, \cdots, \xi^{-1}, \cdots), \quad \xi \in \mu,$$

with ξ and ξ^{-1} at the v-th and the w-th places respectively, and 1 at all other places. For simplicity, we identify the two groups in (3.1).

For each place v, let s_v be the local section defined in Section 1.1. According to [**Sun**, Section 3], there is a section $G(\mathbf{A}) \to \overline{G(\mathbf{A})}$ satisfying the following properties:

(1) $s|_{N(\mathbf{A})} = \prod_v s_v|_{N(F_v)}$ is a homomorphism;
(2) $s|_{T(\mathbf{A})} = \prod_v s_v|_{T(F_v)}$;
(3) $s(w_\sigma) = \prod_v s_v(w_\sigma)$, $\forall \sigma \in \mathfrak{S}_r$;
(4) for any non-archimedean v, $s|_{U_v} = s_v|_{U_v}$ for some compact open subgroup of $\overline{G(\mathbf{A})}$, and $U_v = K_v$ for each $v \notin V$.

By property (1), s is a splitting homomorphism from $N = N(\mathbf{A})$ into $\overline{G}$. We identify N as the subgroup $\overline{G}$ via s.

There is a splitting homomorphism $\psi : G(F) \to \overline{G(\mathbf{A})}$ such that $\psi = s$ on $N(F)$, $T(F)$ and the set $\{w_\sigma : \sigma \in \mathfrak{S}_r\}$. We identify $G(F)$ with a subgroup of $\overline{G(\mathbf{A})}$ via this map.

We now study the induced covering of the diagonal subgroup. By a global Steinberg cocycle, we mean a map $c_{\mathbf{A}} : \mathbf{A}^\times \times \mathbf{A}^\times \to \mu_n$ which is the product of local Steinberg cocycles and is trivial on $F^\times \times F^\times$. This means

$$c_{\mathbf{A}}(x,y) = \prod_v c_v(x_v, y_v), \qquad \forall x, y \in \mathbf{A}^\times;$$

$$c_{\mathbf{A}}(x,y) = 1, \qquad \forall x, y \in F^\times.$$

By [**Moo68**, Theorem 7.1], there is an integer k for each global Steinberg cocycle $c_{\mathbf{A}}$ such that $c_{\mathbf{A}}(x,y) = (x,y)^k_{\mathbf{A}}$, $\forall x, y \in \mathbf{A}^\times$. Here $(x,y)_{\mathbf{A}} = \prod_v (x_v, y_v)_v$ is the global n-th Hilbert symbol.

Fix the global section as above, most local formulas for cocycles are also true in the global case. In particular, corresponding to (1.13) is

$$[s(a), s(b)] = [\iota(\det(a)), \iota(\det(b))] c_{\mathbf{A}}(\det(a), \det(b))^{-1} \prod_i c_{\mathbf{A}}(a_i, b_i), \tag{3.2}$$

where $\iota : \mathbf{A}^\times \to \overline{T_1(\mathbf{A})}$ defined similarly to (1.5).

LEMMA 3.1. *There is a maximal abelian subgroup $\overline{A'}$ of $\overline{T}$ containing group $\mathfrak{Z}(\overline{T})\, T(F)$. Furthermore, both $\mathfrak{Z}(\overline{T})\, T(F) \backslash \overline{T}$ and $\overline{A'} \backslash \overline{T}$ are compact.*

PROOF. Since $\overline{T(\mathbf{A})}$ splits over $T(F)$, $T(F)$ is an abelian subgroup in $\overline{T(\mathbf{A})}$. Extend $T(F)$ to a maximal abelian subgroup $\overline{A'}$ of $\overline{T}$. We get the first part of the lemma.

By (1.15), $\mathfrak{Z}(\overline{T})$ contains $\overline{T^m}$ for some integer m. Observe that $\overline{T(\mathbf{R}^+)}$ is contained in $\overline{T^m}$. So $\mathfrak{Z}(\overline{T})\, T(F) \supset \overline{T(F) T(\mathbf{R}^+)}$ and $\mathfrak{Z}(\overline{T})\, T(F) \backslash \overline{T}$ is compact. □

The following lemma is a generalization of [**KP84**, Lemma II.1.1].

LEMMA 3.2. *Assume the following condition is satisfying by the covering $\overline{G}$:*

$$[\iota(\cdot), \iota(\cdot)] : \mathbf{A}^\times \times \mathbf{A}^\times \to \mu_n \textit{ is a global Steinberg cocycle.} \tag{3.3}$$

Then $\overline{A'} = \mathfrak{Z}(\overline{T})\, T(F)$ and $\mathfrak{Z}(\overline{T}) = \mathfrak{Z}(\overline{G}) \overline{T^n}$.

PROOF. By the assumption and (3.2), we may assume

$$[s(a), s(b)] = c_{\mathbf{A}}(\det(a), \det(b))^m \prod_i c_{\mathbf{A}}(a_i, b_i). \tag{3.4}$$

To show $\mathfrak{Z}(\overline{T}) T(F) \supset \overline{A'}$, let $t \in \overline{A'}$. Applying (3.4) to t and $h_{ij}(a)$ with $a \in F^\times$, we get $c_{\mathbf{A}}(t_i t_j^{-1}, a) = 1$ for any i, j and $a \in F^\times$. It then follows from [**Wei74**, XIII§5, Propostion 8] that $t_i t_j^{-1} \in \mathbf{A}^{\times n} F^\times$. So we may assume $t = s(xI_r)\overline{H(\mathbf{A}^{\times n} F^\times)}$ for $x \in \mathbf{A}^\times$ and only need to show that if $s(xI_r) \in \overline{A'}$ then $s(xI_r) \in \mathfrak{Z}(\overline{T}) T(F)$. Applying (3.4) to $s(xI_r)$ and a for any $a \in T(F)$, we get $c_{\mathbf{A}}(x, \det(a))^{mr+1} = 1$. If l is the greatest common divisor of n and $mr+1$, then $x = yu$ with $y \in \mathbf{A}^{\times n/l}$ and $u \in F^\times$. Observe that $s(yI_r) \in \mathfrak{Z}(\overline{T})$. It follows that $s(xI_r) \in \mathfrak{Z}(\overline{T}) T(F)$.

Similarly, By (3.4), $\mathfrak{Z}(\overline{T}) \supset \mathfrak{Z}(\overline{G}) \overline{T^n}$. And any element in $\mathfrak{Z}(\overline{T})$ must be of the form $s(xI_r) t^n \xi$ for some $x \in \mathbf{A}^\times$, $t \in \overline{T}$ and $\xi \in \mu$. We are then left to show that $s(xI_r) \in \mathfrak{Z}(\overline{T})$ implies $s(xI_r) \in \mathfrak{Z}(\overline{G})$, which is the assertions of Lemma 1.1 in the global case. □

3.2. Representations of Metaplectic Groups

We recall some basic facts on representations of metaplectic groups.

Suppose M is a Levi subgroup of G and for each v, (π_v, H_v) is a representation of $\overline{M_v} = \overline{M(F_v)}$ such that for almost all v, π_v is the Langlands quotient of some induced representation from an irreducible representation of the diagonal subgroup $\overline{T_v}$. Then for almost all v, H_v contains the $(K \cap M)_v^*$-invariant vector v_{ρ_v} (refer to the remark after (1.27)). Define a representation $\pi = \otimes \pi_v$ on H, where H is the Hilbert restricted product of H_v relative to v_{ρ_v} (refer to [**Fla79**]) and

$$\pi(m)x = \otimes_v \pi_v(m_v)x_v, \quad \forall m = \prod_v m_v \in \overline{M(\mathbf{A})}, \; m_v \in \overline{M_v}, \quad x = \otimes x_v \in H.$$

Remark that to form the tensor product, we need v_{ρ_v} to be unimodular. This forces us to choose the measure such that $(K \cap M)_v^*$ has measure one. The following lemma can be proved by the same method as in [**Fla79**, Theorem 3].

LEMMA 3.3. *Any admissible irreducible representation π of $\overline{M_{\mathbf{A}}}$ is of the form $\pi = \otimes \pi_v$ where for each v, π_v is an admissible representation of $\overline{M_v}$.* □

It is not hard to see that

$$i_{\overline{M(\mathbf{A})}}^{\overline{G(\mathbf{A})}} \otimes_v \rho_v = \otimes_v i_{\overline{M_v}}^{\overline{G_v}} \rho_v. \tag{3.5}$$

Let $L^2(\overline{G(F)} \backslash \overline{G(\mathbf{A})})$ be the space of square integrable functions on $\overline{G(F)} \backslash \overline{G(\mathbf{A})}$. The group $\overline{G(\mathbf{A})}$ acts on it by right translation.

For the convenience of reference, (only in this section) we denote by $\overline{{}^{\mu}G}$ the cover of the group $G = G(\mathbf{A})$ by μ. Suppose $j : \mu \to \mathbf{C}^\times$ is a homomorphism. Define

$$L^2\left(\overline{{}^{\mu}G}, j\right) = \left\{ f \in L^2\left(G(F) \backslash \overline{{}^{\mu}G(\mathbf{A})}\right) : f(\xi g) = j(\xi) f(g), \forall \xi \in \mu_n, g \in \overline{{}^{\mu}G(\mathbf{A})} \right\}.$$

Then we have the following trivial decomposition (by the duality on abelian groups):

$$L^2\left(G(F) \backslash \overline{{}^{\mu}G(\mathbf{A})}\right) = \oplus_{j:\mu \to \mathbf{C}^\times} L^2\left(\overline{{}^{\mu}G}, j\right). \tag{3.6}$$

Fix a homomorphism $j : \mu \to \mathbf{C}^\times$. Let ν be the image of j. Then there are a uniquely determined covering $\overline{{}^{\nu}G}$ and a surjective morphism φ such that the diagram

$$\begin{array}{ccccccccc} 1 & \longrightarrow & \mu & \longrightarrow & \overline{{}^{\mu}G} & \longrightarrow & G & \longrightarrow & 1 \\ & & \downarrow{\scriptstyle j} & & \downarrow{\scriptstyle \varphi} & & \| & & \\ 1 & \longrightarrow & \nu & \longrightarrow & \overline{{}^{\nu}G} & \longrightarrow & G & \longrightarrow & 1 \end{array}$$

commutes.

Note ν is a subgroup of $\mathbf{C}^\times$. Let $i : \nu \to \mathbf{C}^\times$ be the identity map. Then we can define a map

$$\varphi^* : L^2\left(\overline{{}^{\nu}G}, i\right) \longrightarrow L^2\left(\overline{{}^{\mu}G}, j\right), \; f \mapsto \varphi^*(f)(g) = f(\varphi(g)).$$

Now choose a map $\psi : \overline{{}^{\nu}G} \to \overline{{}^{\mu}G}$ such that $\varphi\psi = 1_{\overline{{}^{\nu}G}}$, and define

$$\psi^* : L^2\left(\overline{{}^{\mu}G}, j\right) \longrightarrow L^2\left(\overline{{}^{\nu}G}, i\right), \; f \mapsto \psi^*(f)(g) = f(\psi(g)).$$

Observe that

$$\psi(x)\psi(y)\psi(xy)^{-1} \in \mathrm{Ker}\varphi = \mathrm{Ker}j,\ \forall x, y \in \overline{{}^{\nu}G}. \tag{3.7}$$

Now we show that φ^* and ψ^* give isomorphisms between spaces $L^2\left(\overline{{}^{\mu}G}, j\right)$ and $L^2\left(\overline{{}^{\nu}G}, i\right)$ as vector spaces. First, $\varphi\psi = 1$ implies $\psi^* \cdot \varphi^* = 1$. On the other hand, if $f \in L^2\left(\overline{{}^{\mu}G}, j\right)$ and $g \in \overline{{}^{\mu}G}$, since $\psi\left(\varphi\left(g\right)\right)g^{-1} \in \mathrm{Ker}\ \varphi = \mathrm{Ker}\ j$, we have

$$\left(\varphi^* \cdot \psi^*\right)\left(f\right)\left(g\right) = f\left(\psi\left(\varphi\left(g\right)\right)\right) = f\left(g\right).$$

Hence $\varphi^* \cdot \psi^* = 1$.

DEFINITION 3.4. Suppose G_1 and G_2 are two groups and (π_i, V_i) is a representation of G_i $(i = 1, 2)$. Then (G_1, π_1, V_1) and (G_2, π_2, V_2) are said to be equivalent if there are maps of sets $\phi_{12} : G_1 \to G_2$ and $\phi_{21} : G_2 \to G_1$ such that $(\pi_1 \cdot \phi_{21}, V_1)$ and (π_2, V_2) are equivalent representations of G_2 , and such that $(\pi_2 \cdot \phi_{12}, V_2)$ and (π_1, V_1) are equivalent representations of G_1.

It is immediate that if (G_1, π_1, V_1) and (G_2, π_2, V_2) are equivalent, then they have the same decomposition into irreducible representations in the obvious sense. This is the motivation for the above definition.

The purpose of the above calculation is the following lemma.

LEMMA 3.5. *Denote by R the right regular representation of a group. Then the two triples $\left(\overline{{}^{\mu}G}, R, L^2\left(\overline{{}^{\mu}G}, j\right)\right)$ and $\left(\overline{{}^{\nu}G}, R, L^2\left(\overline{{}^{\nu}G}, i\right)\right)$ are equivalent.*

PROOF. We have constructed the maps φ and ψ which give vector space isomorphisms between $L^2\left(\overline{{}^{\mu}G}, j\right)$ and $L^2\left(\overline{{}^{\nu}G}, i\right)$. It is easy to see that $\left(R \cdot \varphi, L^2\left(\overline{{}^{\nu}G}, i\right)\right)$ and $\left(R \cdot \psi, L^2\left(\overline{{}^{\mu}G}, j\right)\right)$ are representations of $\overline{{}^{\mu}G}$ and $\overline{{}^{\nu}G}$ respectively (using (3.7)). So we are left to show that

(1) φ^* intertwines the representations $\left(R \cdot \varphi, L^2\left(\overline{{}^{\nu}G}, i\right)\right)$ and $\left(R, L^2\left(\overline{{}^{\mu}G}, j\right)\right)$;

(2) ψ^* intertwines the representations $\left(R \cdot \psi, L^2\left(\overline{{}^{\mu}G}, j\right)\right)$ and $\left(R, L^2\left(\overline{{}^{\nu}G}, i\right)\right)$.

Proof of (1): Suppose $f \in L^2\left(\overline{{}^{\nu}G}, i\right)$, $g \in \overline{{}^{\mu}G}$. For any $x \in \overline{{}^{\mu}G}$,

$$\varphi^*\left(\left(R \cdot \varphi\right)\left(g\right)f\right)\left(x\right) = \left(R \cdot \varphi\right)\left(g\right)f\left(\varphi\left(x\right)\right) = f\left(\varphi\left(xg\right)\right) = \left(R\left(g\right)\varphi^* f\right)\left(x\right).$$

Proof of (2): Suppose $f \in L^2\left(\overline{{}^{\mu}G}, j\right)$ and $g \in \overline{{}^{\nu}G}$. For any $x \in \overline{{}^{\nu}G}$,

$$\psi^*\left(\left(R \cdot \psi\right)\left(g\right)f\right)\left(x\right) = f\left(\psi\left(x\right)\psi\left(g\right)\right) = f\left(\psi\left(xg\right)\right) = \left(R\left(g\right)\psi^* f\right)\left(x\right).$$

□

For each finite subgroup ν of $\mathbf{C}^\times$, let $e(\nu)$ be the number of total embeddings $\nu \to \mathbf{C}^\times$. By (3.6) and lemma 3.5, we get the following

LEMMA 3.6. *We have the decompostion*

$$L^2\left(G\left(F\right)\backslash\overline{{}^{\mu}G\left(\mathbf{A}\right)}\right) = \oplus_{\nu} e(\nu) L^2\left(\overline{{}^{\nu}G}, i\right)$$

where the sum is over all subgroups ν of $\mathbf{C}^\times$ which are also quotient groups of μ.
□

By the above lemma, we only need to consider the $\overline{{}^{\mu}G\left(\mathbf{A}\right)}$-module $L^2\left(\overline{{}^{\mu}G}, i\right)$ for the decomposition problem. Hence starting from now on, we only consider genuine representations. They are representations π of $\overline{G}$ such that $\pi(\xi) = \xi$ for any $\xi \in \mu$.

Let $\mathfrak{Z}\left(\overline{G}\right)$ be the center of the group $\overline{G(\mathbf{A})}$ and $\mathfrak{Z}\left(\overline{G}\right)^*$ be the group of all genuine unitary characters χ of $\mathfrak{Z}(\overline{G(\mathbf{A})})$. For each $\chi \in \mathfrak{Z}\left(\overline{G}\right)^*$, denote

$$L^2\left(\overline{G},\chi\right) = \left\{f \in L^2\left(\overline{G(F)}\backslash\overline{G(\mathbf{A})}\right) : f(zg) = \chi(z)f(g), \forall z \in \mathfrak{Z}\left(\overline{G}\right), g \in \overline{G(\mathbf{A})}\right\}.$$

Then by harmonic analysis on abelian groups, we have

$$L^2\left(\overline{G},i\right) = \int_{\mathfrak{Z}\left(\overline{G}\right)^*}^{\oplus} L^2\left(\overline{G},\chi\right) d\chi. \tag{3.8}$$

So we need to further decompose the space $L^2\left(\overline{G},\chi\right)$ where $\overline{G}$ is a cover of G by a subgroup μ of $\mathbf{C}^\times$ and χ is a genuine central character.

3.3. Statement of the Problem

Suppose M is a standard Levi subgroup of G corresponding to the partition $r = p_1 + \cdots + p_k$. Let $\overline{M} = \overline{M(\mathbf{A})}$.

For a central character χ_M of $\overline{M}$, denote by $L^2\left(\overline{M},\chi_M\right)$ the space of all functions such that

$$f(zlm) = \chi_M(z) f(m), \quad \forall z \in \mathfrak{Z}\left(\overline{M(\mathbf{A})}\right), m \in \overline{M(\mathbf{A})}, l \in M(F)$$

and that $|f| \in L^2\left(\mathfrak{Z}\left(\overline{M(\mathbf{A})}\right) M(F)\backslash\overline{M(\mathbf{A})}\right)$.

A function on $\overline{M(\mathbf{A})}$ is called cuspidal if for any upper unipotent subgroup N of M,

$$\int_{N(F)\backslash N(\mathbf{A})} f(nm)\, dn = 0, \quad \text{almost everywhere on } \overline{M(\mathbf{A})}. \tag{3.9}$$

Denote by $L_0^2\left(\overline{M},\chi_M\right)$ the subspace of $L^2\left(\overline{M},\chi_M\right)$ consisting of cuspidal forms.

Remark that when $M = T$, (3.9) is vacuous so we use $L^2\left(\overline{A},\chi_T\right)$ instead of $L_0^2\left(\overline{A},\chi_T\right)$. Even in this case, any genuine irreducible representation is an infinite dimensional representation. However in the next section, we are going to show the space $L^2\left(\overline{A},\chi_T\right)$ is a direct sum of at most countably many copies of an irreducible representations.

A slowly increasing smooth function $\phi : M(F)\,N(\mathbf{A})\backslash\overline{G(\mathbf{A})}$ is called automorphic [**MW93**, I.2.17.] if the space generated by $\phi(gk)$, $k \in \overline{K}$ is finite dimensional for any fixed g and the space generated by $\mathfrak{z}\phi$ for $\mathfrak{z}$ running over the center of the universal enveloping algebra of the Lie algebra of $\prod_{v=\infty} G_v$ is finite dimensional. Recall that a smooth $\overline{K}$-finite function ϕ is automorphic if and only if for any $k \in \overline{K}$, $\phi(k)(m) = \Delta(m)^{-1/2}\phi(mk)$ is automorphic over $M(F)\backslash\overline{M(\mathbf{A})}$ [**MW93**, I.2.17.].

An irreducible representation of $\overline{M(\mathbf{A})}$ with central character χ_M is called cuspidal automorphic if it is equivalent to a subspace of $L_0^2\left(\overline{M},\chi_M\right)$ [**BJ79**].

Suppose ρ is the right regular representation on an irreducible space of cuspidal automorphic forms on $M(F)\backslash\overline{M(\mathbf{A})}$. Denote by V_ρ the isotropic subspace of the space of automorphic forms on $M(F)\backslash\overline{M(\mathbf{A})}$, of type ρ. Remark that V_ρ is isomorphic to a direct sum of copies of ρ. And V_ρ is equipped with the natural inner product $\langle\cdot,\cdot\rangle = \langle\cdot,\cdot\rangle_{\overline{M}}$ defined by

$$\langle h',h\rangle = \int_{\mathfrak{Z}(\overline{M(\mathbf{A})})M(F)\backslash\overline{M(\mathbf{A})}} \overline{h'(m)}h(m)\, dm.$$

Remark that if "multiplicity one" is true, then the right regular representation on V_ρ is isomorphic to ρ. For simplicity, we also use ρ to denote the equivalence classes of ρ.

Denote by $X(M)$ the subspace of $\underline{s} \in \mathbf{C}^k$ such that $\sum s_i p_i = 0$. If $\underline{s} \in X(M)$, define $V_\rho[\underline{s}]$ to be the space of cuspidal automorphic forms

$$\left\{ f(m) \,|\det(g_1)|^{s_1} \cdots |\det(g_k)|^{s_k}, m \in \overline{M}, p(m) = \operatorname{diag}(g_1, \cdots, g_k) : f \in V_\rho \right\}.$$

$\overline{M}$ acts on it by the right translation. This representation is equivalent to the representation on the space V_ρ via the action given by

$$\rho[\underline{s}](m) = \rho(m) \,|\det g_1|^{s_1} \cdots |\det g_k|^{s_k}, \text{ for } p(m) = \operatorname{diag}(g_1, \cdots, g_k).$$

Denote by $I(\rho, \underline{s})$ the space of smooth functions f on $\overline{G}$ such that

$$\Delta_P(m)^{-1/2} f(mg) \in V_\rho[\underline{s}], \quad \forall g \in \overline{G}, \forall m \in \overline{M}.$$

This space is equivalent to $i_{\overline{M}}^{\overline{G}} V_\rho[\underline{s}]$. Let $I(\rho) = I(\rho, \underline{0})$. A smooth $\overline{K}$-finite function in $I(\rho, \underline{s})$ can be identified with an automorphic form on $M(F) N(\mathbf{A}) \backslash \overline{G(\mathbf{A})}$. We define a pairing on $I(\rho, -\overline{\underline{s}}) \times I(\rho, \underline{s})$ by

$$\langle f', f \rangle = \langle f', f \rangle_{\overline{G}} = \int_{\overline{M} N \backslash \overline{G}} \langle f'(g), f(g) \rangle_{\overline{M}} \, dg \quad \forall f' \in I(\rho, -\overline{\underline{s}}), \ f \in I(\rho, \underline{s}).$$

This pairing coincides with the inner product in $L^2(\overline{G}, \chi)$ when $\operatorname{Re}(\underline{s}) = \underline{0}$.

Suppose $\sigma \in \mathfrak{S}_k$. We consider it as an element in $\mathfrak{S}_r$ permuting intervals just as we did in the local case. Recall we always identify σ with the element w_σ given by (1.2). We call $\Theta = \left\{ (M, \rho[\underline{s}]) : \underline{s} \in \mathbf{C}^k \right\}$ a cuspidal datum, where M is a standard Levi subgroup of G and ρ is a cuspidal automorphic representation of $\overline{M(\mathbf{A})}$ whose restrictions to $\mathfrak{Z}\left(\overline{G(\mathbf{A})}\right)$ is χ. Two cuspidal data $\Theta = \left\{ (M, \rho[\underline{s}]) : \underline{s} \in \mathbf{C}^k \right\}$ and $\Theta' = \left\{ (M', \rho'[\underline{s}]) : \underline{s} \in \mathbf{C}^k \right\}$ are called equivalent if there is a $\sigma \in \mathfrak{S}_k$ such that $\sigma\Theta = \Theta'$, i.e., $\left\{ (\sigma(M), \sigma(\rho)[\underline{s}]) : \underline{s} \in \mathbf{C}^k \right\} = \left\{ (M', \rho'[\underline{s}]) : \underline{s} \in \mathbf{C}^k \right\}$.

Define the intertwining operator

$$M(\sigma, \rho, \underline{s}) : I(\rho, \underline{s}) \to I(\sigma\rho, \sigma\underline{s}), \quad M(\sigma, \rho, \underline{s})(f)(g) = \int_{N_\sigma} f(\sigma^{-1} n g) \, dn. \tag{3.10}$$

The operator $M(\sigma, \rho, \underline{s})$ can be written as the tensor product of local intertwining operators in the sense of [**MW93**, II 1.9.].

Denote by $\mathbf{I}(\rho)$ the space of holomorphic functions $\phi : X(M) \to I(\rho)$, of Paley-Wiener type and $\overline{K}$-finite. We identify $\phi(\underline{s})$ with an element in $I(\rho, \underline{s})$.

Fix a very large $\underline{\lambda} \in X(M) \cap \mathbf{R}^k$. For $\phi \in \mathbf{I}(\rho), \phi' \in \mathbf{I}(\rho')$ (ρ' is a representation of $\overline{M'(\mathbf{A})}$ defined analogous to ρ),

(3.11)

$$\langle \phi', \phi \rangle = (2\pi)^{1-k} \int_{\underline{s} \in X(M), \operatorname{Re}(\underline{s}) = \underline{\lambda}} \sum_{\sigma \in \mathfrak{S}(\rho, \rho')} \left\langle M\left(\sigma^{-1}, \sigma\rho, -\sigma\overline{\underline{s}}\right) \phi'(-\sigma\overline{\underline{s}}), \phi(\underline{s}) \right\rangle d\underline{s}.$$

where $\mathfrak{S}(\rho, \rho') = \{\sigma \in \mathfrak{S}_r : \ \sigma(M) = M', \sigma\rho = \rho'\}$. The integral does not depend on the choice of $\underline{\lambda}$.

The Eisenstein series is defined as

$$E(g, \phi, \underline{s}) = \sum_{\gamma \in P(F) \backslash G(F)} \phi(\gamma g, \underline{s}), \quad \forall \phi \in \mathbf{I}(\rho), \ g \in \overline{G(\mathbf{A})}, \ \underline{s} \in X(M).$$

The sum converges for $\underline{s}$ large, it then can be continued to a meromorphic function on $X(M)$.

For each ϕ as above, denote by $\mathfrak{F}\phi$ the Fourier transform of $E(g,\phi,\underline{s})$ in the sense of Langlands (see [**Art79**, p257]). According to the theory of Eisenstein series [**Lan76**, lemma 4.6], $\mathfrak{F}$ intertwines $\mathbf{I}(\rho)$ and the subspace

$$L^2\left(\overline{G},\chi\right)_\Theta = \text{the completion of } \int_{\mathrm{Re}(\underline{s})=\underline{\lambda}} E\left(i_{\overline{M}}^{\overline{G}} V_\rho, \underline{s}\right) d\underline{s}$$

such that $\langle\phi',\phi\rangle = \langle\mathfrak{F}\phi',\mathfrak{F}\phi\rangle$, where Θ is the cuspidal datum that ϕ belongs to. Denote $\{\Theta\} = \{\sigma\Theta : \sigma\in\mathfrak{S}_k\}$, the conjugate class containing Θ. Denote $L^2\left(\overline{G},\chi\right)_{\{\Theta\}}$ the span of $L^2\left(\overline{G},\chi\right)_{\sigma\Theta}$, $\sigma\in\mathfrak{S}_k$. Then $L^2\left(\overline{G},\chi\right) = \oplus_{\{\Theta\}} L^2\left(\overline{G},\chi\right)_{\{\Theta\}}$. In other words, the space $L^2\left(\overline{G},\chi\right)$ is the completion of the space generated by $\mathfrak{F}\phi$ where ϕ runs over $\mathbf{I}(\rho)$ and (M,ρ) runs over the set of cuspidal data.

Summing up, we have

$$L^2\left(\overline{G},\chi\right) = \oplus_{\{\Theta\}} L^2\left(\overline{G},\chi\right)_{\{\Theta\}}. \tag{3.12}$$

In this paper, $\oplus$ is the completion after the sum. For the precise statement and the proof of the last identity, refer to [**Art79**, lemma 6], [**Lan76**, lemma 4.6], or [**MW93**, p. 108].

To further decompose the space $L^2\left(\overline{G},j\right)_{\{\Theta\}}$, we need to move the domain of integration in (3.11). We consider the Borel case, which will be solved in Section 3.5. More explicitly, we denote by Θ_0 a cuspidal datum whose first component is the diagonal subgroup T, i.e., $\Theta_0 = (T,\rho[\underline{s}]) : \underline{s}\in\mathbf{C}^r$ for an automorphic irreducible representation ρ of $\overline{T}$. Section 3.5 will decompose $L^2\left(\overline{G},\chi\right)_{\{\Theta_0\}}$, which is the main result in this paper.

3.4. Representations of the Diagonal Subgroup

Since $\overline{G}$ splits over K_v for almost all v, $\overline{T}$ splits over $T(\mathfrak{O}_v)$ for almost all v. This implies that $\overline{T(\mathfrak{O}_v)}$ is commutative for almost all v. Fix a maximal abelian subgroup $\overline{A_v}$ of $\overline{T_v}$ for each v, such that for almost all $T(\mathfrak{O}_v)\subset A_v$. Now let

$$\overline{A} = \{a\in\overline{T} : a_v\in\overline{A_v}, \forall v\}.$$

Then by the above remarks, we get

LEMMA 3.7. *The subgroup $\overline{A}$ is open in $\overline{T}$ and $\overline{A}\backslash\overline{T}$ is a discrete abelian group which is countable.* □

Suppose ρ is a quasi-character of $\overline{A}$. Define $i_{\overline{A}}^{\overline{T}}\rho$ to be the compact induction. The space of $i_{\overline{A}}^{\overline{T}}\rho$ consists of all functions $f:\overline{T}\to\mathbf{C}$ such that

(1) $f(at) = \rho(a) f(t)$, $\forall a\in\overline{A}$, $t\in\overline{T}$;

(2) f is zero outside finitely many cosets in $\overline{A}\backslash\overline{T}$.

The group $\overline{T}$ acts on this space by right translation. The above conditions imply that f is smooth on $\overline{T}$. The definition coincides with the one defined in [**KP84**, p.54-p.55].

Assume ρ to be unitary. We can give a pre-Hilbert space structure to $i_{\overline{A}}^{\overline{T}}\rho$ by installing the following inner product:

$$\langle f,f'\rangle = \sum_{\gamma\in\overline{A}\backslash\overline{T}} \overline{f}(\gamma) f'(\gamma) \quad \forall f,f'\in i_{\overline{A}}^{\overline{T}}\rho.$$

Denote by $L^2\left(\overline{T}, \rho\right)$ the space of functions on the group $\overline{T(\mathbf{A})}$ such that $f(at) = \rho(a) f(t)$, $\forall a \in \overline{A}, t \in \overline{M}$, and $|f| \in L^2\left(\overline{A}\backslash\overline{T}\right)$.

LEMMA 3.8. *As $\overline{T}$-spaces,*

$$L^2\left(\overline{T}, \rho\right) = \text{ the completion of } i_{\overline{A}}^{\overline{T}}\rho = \text{ the completion of } \otimes_v\, i_{\overline{A}_v}^{\overline{T}_v}\rho_v.$$

In particular, $i_{\overline{A}}^{\overline{T}}\rho$ is irreducible. □

By this lemma, the representation on the space $L^2\left(\overline{A}, \rho\right)$ can be written as a tensor product of local induced representations. But it is not $T(F)$-invariant. We now construct an isomorphism between this space and some $T(F)$-invariant space.

Fix a maximal abelian subgroup $\overline{A'}$ of $\overline{T}$ containing $\mathfrak{Z}(\overline{T})T(F)$ (see Lemma 3.1). Suppose χ is a unitary character of $\overline{A'}$ trivial on $A(F)$. We can define the induced representation $i_{\overline{A'}}^{\overline{T}}\chi$ to be the space of all smooth functions f on $\overline{T}$ such that $f(at) = \chi(a) f(t)$ for any $a \in \overline{A'}$ and $t \in \overline{T}$. The induced representation $i_{\overline{A'}}^{\overline{T}}\chi$ is irreducible by Mackey's criterion.

Suppose $\rho = \chi$ on $\overline{A'} \cap \overline{A}$, define (refer to [**KP84**, p. 55])

$$\Psi : i_{\overline{A}}^{\overline{T}}\rho \to i_{\overline{A'}}^{\overline{T}}\chi, \qquad \Psi(f)(x) = \sum_{a \in \left(\overline{A'} \cap \overline{A}\right)\backslash\overline{A'}} \chi^{-1}(a) f(ax). \tag{3.13}$$

Note the sum is finite. We claim Ψ preserves the inner product. Indeed, let $f, g \in i_{\overline{A}}^{\overline{T}}\rho$. Then

$$\int_{\overline{A'}\backslash\overline{T}} dx \Psi(f)(x)\overline{\Psi(g)(x)}$$

$$= \int_{\overline{A'}\backslash\overline{T}} dx \sum_{a,b \in (\overline{A'}\cap\overline{A})\backslash\overline{A'}} \chi(a^{-1}b) f(ax)\overline{g(bx)}$$

$$= \int_{\overline{A'}\backslash\overline{T}} dx \sum_{a,c \in (\overline{A'}\cap\overline{A})\backslash\overline{A'}} \chi(c) f(ax)\overline{g(cax)} \tag{3.14}$$

$$= \int_{(\overline{A'}\cap\overline{A})\backslash\overline{T}} dy \sum_{c \in (\overline{A'}\cap\overline{A})\backslash\overline{A'}} \chi(c) f(y)\overline{g(cy)} \tag{3.15}$$

$$= \sum_{\gamma \in \overline{A}\backslash\overline{T}} \int_{(\overline{A'}\cap\overline{A})\backslash\overline{A}} dx \sum_{c \in (\overline{A'}\cap\overline{A})\backslash\overline{A'}} \chi(c) f(x\gamma)\overline{g(cx\gamma)} \tag{3.16}$$

$$= \sum_{\gamma \in \overline{A}\backslash\overline{T}} \sum_{c \in (\overline{A'}\cap\overline{A})\backslash\overline{A'}} \chi(c) f(\gamma)\overline{g(c\gamma)} \int_{(\overline{A'}\cap\overline{A})\backslash\overline{A}} [c, x] dx \tag{3.17}$$

$$= \sum_{\gamma \in \overline{A}\backslash\overline{T}} f(\gamma)\overline{g(\gamma)} \int_{(\overline{A'}\cap\overline{A})\backslash\overline{A}} dx$$

We made the following changes of variables: $b = ca$ to get (3.14); $y = ax$ to get (3.15); $y = x\gamma$ to get (3.16). We also observed that in (3.17), the integral is 0 unless $c = 1$.

Observe that both $i_{\overline{A}}^{\overline{T}}\rho$ and $i_{\overline{A'}}^{\overline{T}}\chi$ are irreducible. So Ψ intertwines them isomorphically which preserves the inner products up to a scalar. So we get as $\overline{T}$-spaces,

$$\text{the completion of } i_{\overline{A}}^{\overline{T}}\rho \cong \text{ the completion of } i_{\overline{A'}}^{\overline{T}}\chi. \tag{3.18}$$

As before, for a unitary character χ_T of $\overline{T}$ trivial on $\overline{T} \cap T(F)$, define

$$(3.19) \quad L^2\left(\overline{T}, \chi_T\right) = \left\{ f : T(F) \backslash \overline{T(\mathbf{A})} \to \mathbf{C} : |f| \in L^2\left(T(F)\mathfrak{Z}\left(\overline{T(\mathbf{A})}\right) \backslash \overline{T(\mathbf{A})}\right), \right.$$
$$\left. f(zt) = \chi_T(z) f(t), \forall z \in \mathfrak{Z}\left(\overline{T(\mathbf{A})}\right), t \in \overline{T(\mathbf{A})} \right\}.$$

LEMMA 3.9. *(1)Two unitary irreducible admissible representations of $\overline{T}$ are equivalent if and only if they have the same central character.*

(2) The $\overline{T}$-space $L^2\left(\overline{T}, \chi_T\right)$ is a direct sum of at most countably many copies of the irreducible representation of $\overline{T(\mathbf{A})}$ whose central character is χ_T. The multiplicity equals the order of the group $(\mathfrak{Z}(\overline{T})T(F)\backslash\overline{A'})^*$.

(3) Any unitary irreducible admissible representation of $\overline{T}$ trivial on $\mathfrak{Z}\left(\overline{T(\mathbf{A})}\right) \cap T(F)$ occurs in $L^2\left(\overline{T}, \chi_T\right)$ for some χ_T.

PROOF. (1) Suppose ρ is a unitary irreducible admissible representations of $\overline{T}$ with central character χ_T. Write $\rho = \otimes_v \rho_v$. By the local Lemma 1.3, for each v, ρ_v is uniquely determined by its central character, hence so is ρ.

(2) Fix a character χ of $\overline{A'}$ extending the central character χ_T. Let

$$(T(F)\mathfrak{Z}(\overline{T})\backslash\overline{A'})^*$$

be the set of unitary characters of $\overline{A'}$ trivial on $T(F)\mathfrak{Z}(\overline{T})$. Then

$$\{\delta\chi : \delta \in (T(F)\mathfrak{Z}(\overline{T})\backslash\overline{A'})^*\}$$

is the set of all characters on $\overline{A'}$ extending χ_T. Define the Fourier transform

$$\hat{f}(t)(\delta) = \int_{\overline{A'}} f(at)\overline{(\delta\chi)(a)}da, \quad \forall t \in \overline{T}, \delta \in \left(T(F)\mathfrak{Z}(\overline{T})\backslash\overline{A'}\right)^*.$$

By the inversion formula, after a suitable choice of measures,

$$f(t) = \int_{(T(F)\mathfrak{Z}(\overline{T})\backslash\overline{A'})^*} \hat{f}(t)(\delta)d\delta.$$

The Plancherel formula tells us that

$$(3.20) \qquad L^2(\overline{T}, \chi_T) = \int_{(T(F)\mathfrak{Z}(\overline{T})\backslash\overline{A'})^*} L^2(\overline{T}, A', \chi\delta)d\delta,$$

where $L^2(\overline{T}, A', \chi\delta) = \{f : \overline{T} \to \mathbf{C} : |f| \in L^2(\overline{A'}\backslash\overline{T}), f(at) = (\delta\chi)(a)f(t), \forall a \in \overline{A'}, t \in \overline{T}\}$.

Since $T(F)\mathfrak{Z}(\overline{T})\backslash\overline{A'}$ is compact, the integral in (3.20) is really a discrete sum of at most countably many terms. Also the integrand in (3.20) is irreducible.

(3) If π is a representation satisfying the condition, then π can be written as a tensor product $\pi = \otimes_v \pi_v$. By Lemma 1.3, each π_v is of the form $\pi_v = i_{\overline{A_v}}^{\overline{T_v}} \rho_v$ for some character ρ_v of $\overline{A_v}$ for each v. We can alway choose ρ_v such that for almost all places, ρ_v is unramified. Then $\pi = i_{\overline{A}}^{\overline{T}} \rho^\circ$ for $\rho = \otimes_v \rho_v$. It follows from (3.18) that $\pi = i_{\overline{A'}}^{\overline{T}} \chi$ for some character χ of $\overline{A'}$. By part (1), we may assume χ is trivial on $T(F)$. □

It follows from Lemma 3.9 and Lemma 3.2 that we have:

COROLLARY 3.10. *If condition (3.3) is satisfied, then* $L^2\left(\overline{T}, \chi_T\right)$ *is irreducible.*
□

We can use Lemma 3.2 to give an explicit description of the set of automorphic representations on $\overline{T}$ in this special case.

COROLLARY 3.11. *Assume condition (3.3) is satisfied. Let* χ *be a unitary character of* $\mathfrak{Z}(\overline{G})$. *There is a one to one correspondence between the set of all automorphic representations on* $\overline{T}$ *whose restriction to* $\mathfrak{Z}\left(\overline{G}\right)$ *is* χ *and the set of all r-tuples of unitary characters* $(\rho_1, \cdots, \rho_r)$ *of* $\mathbf{A}^{\times n}/F^{\times n}$ *such that* $\rho_1 \cdots \rho_r = \chi$ *on* $\mathbf{A}^{\times n}$. *Specifically, let* ρ_i *be as above. Let* $\rho = \rho_1 \otimes \cdots \otimes \rho_r$ *be tensor product representation of* $\overline{T^n}$. *Extend* ρ *to a character* χ_T *of* $\mathfrak{Z}(\overline{T}) = \mathfrak{Z}(\overline{G})\overline{T^n}$ *by* χ. *Then the space* $L^2(\overline{T}, \chi)$ *is irreducible space of* $\overline{T}$. □

3.5. The Main Theorem

Assume F is a number field satisfying Assumption 0.1, i.e., $(-1,-1)_v = 1$ for any place v. Let $\overline{G}$ is a metaplectic cover of G such that $[\overline{H(\mathbf{A})}, \overline{H(\mathbf{A})}] = \mu_n$.

Recall the definitions of the global L-function and ε-factor. Suppose μ is a quasi-character of $\mathbf{A}^\times/F^\times$, then

$$L\left(s, \mu\right) = \prod_v L_v\left(s, \mu_v\right); \quad \varepsilon\left(s, \mu\right) = \prod_v \varepsilon_v\left(s, \mu_v\right).$$

The ε-factor is a function holomorphic and non-vanishing everywhere in $\mathbf{C}$. Suppose μ is unitary. If $\mu = 1$, then $L\left(s, \mu\right)$ has a simple pole at $s = 0$ or 1 and is analytic elsewhere. Otherwise, it is analytic everywhere. Furthermore we have the following functional equation:

$$L\left(s, \mu\right) = \varepsilon\left(s, \mu\right) L\left(1 - s, \overline{\mu}\right). \tag{3.21}$$

Let ρ be an irreducible representation of $\overline{T}$ with central character χ_ρ. Recall $h_{ij}(x^n) \in \mathfrak{Z}(\overline{T})$ for any $x \in \mathbf{A}$. Define a character ρ^n_{ij} of $\mathbf{A}$ by $\rho^n_{ij}(x) = \chi_\rho(h_{ij}(x^n))$.

For $\sigma \in \mathfrak{S}_r$, $\underline{s} \in \mathbf{C}^r$, define $r\left(\sigma, \rho, \underline{s}\right) = \prod_v r_v\left(\sigma, \rho_v, \underline{s}\right)$ See (1.46) for the definition of r_v. Then by proposition 1.17,

$$N\left(\sigma, \rho, \underline{s}\right) = r\left(\sigma, \rho, \underline{s}\right)^{-1} M\left(\sigma, \rho, \underline{s}\right)$$

is normalized. This means that the operator $N\left(\sigma, \rho, \underline{s}\right)$

- is holomorphic in $\{\underline{s} \in \mathbf{C}^r : \operatorname{Re}(s_i) \geq \operatorname{Re}(s_j), i < j\}$;
- can be written as a tensor product $\otimes_v N_v\left(\sigma, \rho, \underline{s}\right)$, i.e., for almost all places v, $N_v\left(\sigma, \rho, \underline{s}\right)$ sends the canonical K_v^*-invariant vector to the canonical K_v^*-invariant vector;
- is such that for any place v, $N_v\left(\tau\sigma, \rho, \underline{s}\right) = N_v\left(\tau, \sigma\rho, \sigma\underline{s}\right) N_v\left(\sigma, \rho, \underline{s}\right)$;
- is unitary when ρ is unitary and $\underline{s} \in i\mathbf{R}^r$.

Since $\rho(-1) = 1$, $\prod_v \rho_v(-1) = 1$. So $\Pi_v \gamma((\rho_v)_{ij}) = \pm 1$ (see page 16). Also $\prod_v |n|_v = 1$. By (1.46) we have

$$r\left(\sigma, \rho, \underline{s}\right) = \pm \prod_{i,j \in \operatorname{inv}(\sigma)} \frac{L\left(n\left(s_i - s_j\right), \rho^n_{ij}\right)}{L\left(n\left(s_i - s_j\right) + 1, \rho^n_{ij}\right) \varepsilon\left(n\left(s_i - s_j\right), \rho^n_{ij}\right)}. \tag{3.22}$$

By properties of the L function, $r\left(\sigma, \rho, \underline{s}\right)$ has a pole when $\rho^n_{ij} = 1$ and $n\left(s_i - s_j\right) = 1$ for $(i, j) \in \operatorname{inv}\left(\sigma\right)$.

Suppose $f \in I(\rho, \underline{s})$, $f' \in I(\sigma\rho, -\sigma\overline{\underline{s}})$. We have the following formulas for the adjoint

$$\begin{aligned} \langle f', M(\sigma, \rho, \underline{s}) f \rangle &= \left\langle M\left(\sigma^{-1}, \sigma\rho, -\sigma\overline{\underline{s}}\right) f', f \right\rangle, \\ r(\sigma, \rho, \underline{s}) &= \overline{r(\sigma^{-1}, \sigma\rho, -\sigma\overline{\underline{s}})}, \\ \langle f', N(\sigma, \rho, \underline{s}) f \rangle &= \left\langle N\left(\sigma^{-1}, \sigma\rho, -\sigma\overline{\underline{s}}\right) f', f \right\rangle. \end{aligned} \tag{3.23}$$

The following lemma was stated for $GL(2)$ in [**Fli80**, p. 159]. We give a proof for completeness.

LEMMA 3.12. *Suppose $r \geq 3$. If $\sigma = (ij)$ is a simple reflection, ρ is a unitary irreducible representation of $\overline{T}$ such that $(ij)(\rho) = \rho$, then*

$$M(\sigma, \rho, \underline{0}) = -1 .$$

PROOF. By the remark at beginning of section 1.6, we can assume that for any v, $\rho^{\circ}(h_{ij}(-1)) = 1$. By the definition of $\gamma(\rho_{ij})$, the sign on the right hand side of (3.22) is positive. It follows from (3.21) that ($\underline{s} \in \mathbf{C}^r$)

$$\lim_{\underline{s} \to \underline{0}} r(\sigma, \rho, \underline{s}) = -1.$$

So we need to show that $N(\sigma, \rho, \underline{0}) = 1$. This follows from the local Proposition 1.18. □

From now on, if the context is clear, we shall drop the letter ρ, for example: $M(\sigma, \underline{s}) = M(\sigma, \rho, \underline{s})$.

Denote by $\mathbf{P}$ the set of all partitions of r. Suppose $p_1 + \cdots + p_m = r$. Denote $\underline{p} = (p_1, \cdots, p_m) \in \mathbf{P}$ and define $d(\underline{p}) = m$. Once we write $\pi_{\underline{p}}$, we mean an equivalence class of irreducible unitary representations which is invariant under $\mathfrak{S}(\underline{p})$, i.e., $\sigma\pi_{\underline{p}} = \pi_{\underline{p}}, \forall\sigma \in \mathfrak{S}(\underline{p})$. We use the following notation in [**MW89**, p. 645-p. 646] (with some modifications):

- $M = M_{\underline{p}} = GL(p_1) \times \cdots \times GL(p_m)$;
- $p'_i = \sum_{k=1}^{i-1} p_k \quad i = 1, \cdots, m+1$;
- $\Delta_i = \left\{p'_i + 1, \cdots, p'_{i+1}\right\} \subset \mathbf{N}$;
- $X(T) = \{\underline{s} \in \mathbf{C}^r : \sum s_i = 0\}$;
- $V(\underline{p}) = \left\{\underline{s} \in X(T) : \ s_i - s_{i+1} = \frac{1}{n} \text{ if } i, i+1 \text{ are in the same interval } \Delta_k\right\}$;
- $V^0(\underline{p}) = \{\underline{s} \in X(T) : \ s_i - s_{i+1} = 0 \text{ if } i, i+1 \text{ are in the same interval } \Delta_k\}$;
- $\lambda(\underline{p}) = \left(\frac{p_1-1}{2n}, \frac{p_1-3}{2n}, \cdots, \frac{1-p_1}{2n}, \frac{p_2-1}{2n}, \cdots, \frac{1-p_m}{2n}\right)$;
- $\mathfrak{S}(\uparrow, \underline{p})$ = the set of all $\sigma \in W$ increasing on each interval Δ_k, i.e.

$$\sigma(i) < \sigma(j), \quad \forall k, \forall i, j \in \Delta_k, i < j;$$

- $w_{\underline{p}}$ =the element in $\mathfrak{S}_r$ which reverses the order on each interval, i.e.,

$$w_{\underline{p}}(p'_k + i) = p'_{k+1} + 1 - i, \quad i = 1, \cdots, p_k.$$

Suppose L is a large real number and D is a subset of $\mathbf{C}^r$ provided with a measure. Denote by $\int_D^L$ the integral on $D \cap \{\underline{s} \in \mathbf{C}^r : \ \|\mathrm{Im}(\underline{s})\| \leq 2L\}$.

Suppose $\phi, \phi' \in \mathbf{I}(\rho)$ as in section 3.3 and $a(\phi, \phi')$, $b(\phi, \phi')$ two linear forms depending on ϕ and ϕ'. We write

$$a(\phi, \phi') =_L b(\phi, \phi') \tag{3.24}$$

if $a(\phi,\phi') - b(\phi,\phi')$ is of the form:

$$\int_D^L c(\underline{s})\,\phi(\underline{s})\,\overline{\phi'(\gamma\overline{\underline{s}})}d\underline{s}$$

where D is a union of Borel subsets in subspaces of $\mathbf{C}^r$ provided with the measure which is a linear combination of the Lebesgue measures on subspaces of $\mathbf{C}^r$, and where $c(\cdot)$ is the restriction of a holomorphic function to D, $\gamma \in GL(r,\mathbf{C})$ such that for $\underline{s} \in D$, we have $\|\mathrm{Im}(\underline{s})\| > L$, $\|\mathrm{Im}(\gamma\underline{s})\| > L$ and $\mathrm{Re}(\underline{s})$ as well as $\mathrm{Re}\gamma(\underline{s})$ is in a compact set independent of L. The convergence of the above integral is assumed whence we write (3.24).

Put

$$f_{\underline{p}}(\underline{s}) = \prod_{k=1}^{m} \prod_{p'_k<i<p'_{k+1}} \left(s_i - s_{i+1} - \frac{1}{n}\right). \tag{3.25}$$

By the properties of the L-function $f_{\underline{p}}(\underline{s})\,r\left(w_{\underline{p}},\pi_{\underline{p}},\underline{s}\right)$ is holomorphic and is a constant on $V(\underline{p})$. We denote this constant by $c'_{\underline{p}}$ and set

$$c_{\underline{p}} = (1/m!)\,(2\pi)^{-d(\underline{p})}\,c'_{\underline{p}}\,.$$

Denote by $(\underline{p},\pi)$ a partition $\underline{p}$ of r together with a unitary irreducible representation $\pi_{\underline{p}}$ as above. Define

$$\langle\phi',\phi\rangle^L_{(\underline{p},\pi)} = c_{\underline{p}} \int^L_{\underline{s}\in V(\underline{p}),\mathrm{Re}(\underline{s})=\lambda(\underline{p})} \sum_{\sigma,\tau}$$

$$\left\langle N\left(w_{\underline{p}},-w_{\underline{p}}\overline{\underline{s}}\right) M\left(\tau^{-1},-\tau w_{\underline{p}}\overline{\underline{s}}\right)\phi'\left(-\tau w_{\underline{p}}\overline{\underline{s}}\right), M\left(\sigma^{-1},\sigma\underline{s}\right)\phi\left(\sigma\underline{s}\right)\right\rangle d_{\underline{p}}\underline{s}\,, \tag{3.26}$$

summing over $\sigma \in \mathfrak{S}(\uparrow,\underline{p}) \cap \mathfrak{S}\left(\pi_{\underline{p}},\rho\right)$, $\tau \in \mathfrak{S}(\uparrow,\underline{p}) \cap \mathfrak{S}\left(\pi_{\underline{p}},\rho'\right)$. Remark that in the next chapter, we are going to show that the right hand side of the above identity is well-defined (lemma 4.5).

We say two irreducible representation ρ and ρ' of $\overline{T}$ are "equivalent upto conjugates and twists" if there are $\underline{s} \in \mathbf{C}^r$ and $\sigma \in \mathfrak{S}_r$ such that $\rho' = \sigma(\rho[\underline{s}])$. Suppose ϱ is an "equivalence class upto conjugates and twists" of unitary irreducible representations of $\overline{T}$. Denote by $\mathbf{P}(\varrho)$ the set of all pairs $(\underline{p},\pi)$, where $\underline{p} \in \mathbf{P}$ is a partition of N and $\pi = \pi_{\underline{p}}$ such that $\pi_{\underline{p}} \in \varrho$.

In the next chapter, we shall show

LEMMA 3.13. (refer to [**MW89**, p. 647])
Suppose $\phi \in \mathbf{I}(\rho)$, $\phi' \in \mathbf{I}(\rho')$ and $T > 0$. We have the following relation:

$$\langle\phi',\phi\rangle =_L \sum_{(\underline{p},\pi)\in\mathbf{P}(\varrho)} \langle\phi',\phi\rangle^L_{(\underline{p},\pi)}\,. \tag{3.27}$$

We fix a ϱ in the remaining part of the section. Suppose $(\underline{p},\pi) \in \mathbf{P}(\varrho)$, $J_{\underline{p}} = J_{(\underline{p},\pi)}$ is a representation of $\overline{M(\mathbf{A})}$ such that at each place v of F, the local component $J_{\underline{p},v}$ is

$$J\left((1-p_1)/2n,(p_1-1)/2n\right) \otimes_{\pi_{\underline{p},v}} \cdots \otimes J\left((1-p_m)/2n,(p_m-1)/2n\right).$$

By the local results, it is immediate that

LEMMA 3.14. *The representation* $J_{(\underline{p},\pi)}$ *is the unique irreducible quotient of* $i_T^M \pi_{\underline{p}}[\lambda(\underline{p})]$. *It is also the image of* $N\left(w_{\underline{p}}, \pi_{\underline{p}}\right)$. □

For $\underline{s} \in V(\underline{p})$, denote by $z(\underline{s}) \in \mathbf{C}^m$ the point such that $z(\underline{s})_k = s_j - \lambda(\underline{p})_j$ for $j \in \Delta_k$. Remark that the normalization of intertwining operators (see page 45) implies that for almost all places v, $N_v\left(w_{\underline{p}}, \underline{s}\right)$ is the identity on a K_v^*-invariant vector of $I_v\left(\pi_{\underline{p},v}, \underline{s}\right)$.

For $\underline{s} \in V(\underline{p})$, let $I\left(J_{(\underline{p},\pi)}, z(\underline{s})\right)$ be the image of the induced representation $I\left(\pi_{\underline{p}}, \underline{s}\right)$ under the map $N\left(w_{\underline{p}}, \underline{s}\right)$. Let $m(\overline{T})$ be the cardinality of the group $(T(F)\mathfrak{Z}(\overline{T})\backslash\overline{A'})^*$. Then $m(\overline{T})$ is at most countable (Lemma 3.9) and we have

$$I\left(J_{(\underline{p},\pi)}, z(\underline{s})\right) = m(\overline{T}) \cdot i_M^{\overline{G}} J_{(\underline{p},\pi)}[z(\underline{s})]. \tag{3.28}$$

Indeed, by Lemma 3.9,

$$I\left(\pi_{\underline{p}}, \underline{s}\right) = i_{\overline{T}}^{\overline{G}} V_\rho[\underline{s}] = m(\overline{T}) \cdot i_{\overline{T}}^{\overline{G}} \rho[\underline{s}].$$

Now apply $N(w_{\underline{p}}, \pi_{\underline{p}})$ to the above isomorphisms, we get (3.28) by Lemma 3.14.

Besides we have:

$$z\left(-w_{\underline{p}}\overline{\underline{s}}\right) = -\overline{z(\underline{s})}\ .$$

The adjoint operator of $N\left(w_{\underline{p}}, \underline{s}\right)$ is $N\left(w_{\underline{p}}, -w_{\underline{p}}\overline{\underline{s}}\right)$. So we can define a $\overline{G}$-invariant sesquilinear form, denoted by $\langle\langle\cdot,\cdot\rangle\rangle$, on $I\left(J_{(\underline{p},\pi)}, -z(\overline{\underline{s}})\right) \times I\left(J_{(\underline{p},\pi)}, z(\underline{s})\right)$ by

$$\left\langle\left\langle N\left(w_{\underline{p}}, -w_{\underline{p}}\overline{\underline{s}}\right) f', N\left(w_{\underline{p}}, \underline{s}\right) f\right\rangle\right\rangle = \left\langle N\left(w_{\underline{p}}, -w_{\underline{p}}\overline{\underline{s}}\right) f', f\right\rangle\ , \tag{3.29}$$

for $f \in I\left(\pi_{\underline{p}}, \underline{s}\right), f' \in I\left(\pi_{\underline{p}}, -w_{\underline{p}}\overline{\underline{s}}\right)$. When $\mathrm{Re}(\underline{s}) = \lambda(\underline{p})$, we have $-w_{\underline{p}}\overline{\underline{s}} = \lambda(\underline{p})$. The above form becomes a hermitian form on $I\left(J_{(\underline{p},\pi)}, z(\underline{s})\right)$. Let

$$\begin{aligned} Z(\underline{p}) &= \left\{z(\underline{s}) : \underline{s} \in V(\underline{p}), \mathrm{Re}(\underline{s}) = \lambda(\underline{p})\right\} \\ &\cong Z_0(\underline{p}) = \left\{\underline{t} \in i\mathbf{R}^m : \sum p_i t_i = 0\right\}, \end{aligned} \tag{3.30}$$

provided with the measure $d_{\underline{p}}\underline{z}$ induced from $d_{\underline{p}}\underline{s}$.

Suppose $\phi \in I(\rho)$. We define a function, denoted by $R_{(\underline{p},\pi)}\phi$, from $Z(\underline{p})$ into $I\left(J_{\underline{p}}\right)$ by

$$\left(R_{(\underline{p},\pi)}\phi\right)(z(\underline{s})) = \sum N\left(w_{\underline{p}}, \underline{s}\right) M\left(\sigma^{-1}, \sigma\underline{s}\right) \phi(\sigma\underline{s})$$

summing over $\sigma \in \mathfrak{S}(\uparrow, \underline{p}) \cap \mathfrak{S}\left(\pi_{\underline{p}}, \rho\right)$. We thus have for $\phi' \in I(\rho')$, $\phi \in I(\rho)$:

$$\langle\phi', \phi\rangle^L_{(\underline{p},\pi)} = c_{\underline{p}} \int^L_{Z(\underline{p})} \left\langle\left\langle R_{\underline{p},\pi}\phi'(\underline{z}), R_{\underline{p},\pi}\phi(\underline{z})\right\rangle\right\rangle d_{\underline{p}}\underline{z}\ . \tag{3.31}$$

In the remaining part of this section let $\tau \in \mathfrak{S}_r$ be an element which permutes the intervals $\{\Delta_k\}$ and is increasing on each of them. Then $(ij)\tau(\pi) = \tau(\pi)$ for any k and $i, j \in \Delta_k$. We also identify a permutation in $\mathfrak{S}_m$ as such an element. We define the element $\tau\underline{p} \in \mathbf{P}$, the map $\tau : Z(\underline{p}) \to Z(\tau\underline{p})$ and the operator $M\left(\tau, J_{(\underline{p},\pi)}, \underline{z}\right)$.

We see that the functions $\Phi_{(\underline{p},\pi)} = R_{(\underline{p},\pi)}\phi$ and $\Phi_{(\tau\underline{p},\tau\pi)} = R_{(\tau\underline{p},\tau\pi)}\phi$ satisfy the condition

$$\Phi_{(\tau\underline{p},\tau\pi)}(\tau\underline{z}) = M\left(\tau, J_{(\underline{p},\pi)}, \underline{z}\right)\Phi_{(\underline{p},\pi)}(\underline{z}) \quad \text{for all } \underline{z} \in Z(\underline{p}) . \tag{3.32}$$

Suppose $\underline{z} \in Z(\underline{p})$, $\Phi, \Phi' \in I\left(J_{(\underline{p},\pi)}, \underline{z}\right)$. We have

$$\left\langle\left\langle M\left(\tau, J_{(\underline{p},\pi)}, \underline{z}\right)\Phi', M\left(\tau, J_{(\underline{p},\pi)}, \underline{z}\right)\Phi\right\rangle\right\rangle = \langle\langle\Phi', \Phi\rangle\rangle . \tag{3.33}$$

See [**MW89**, p. 648] for the proof.

For $(\underline{p},\pi) \in \mathbf{P}(\varrho)$, introduce the set with multiplicities $\{(\underline{p},\pi)\}$, which we call the class of $(\underline{p},\pi)$. It is formed by $(\tau\underline{p},\tau\pi)$ for τ verifying the above hypothesis. The multiplicity associated to $(\underline{p},\pi)$ is the number of such τ fixing it. Denote by $\mathbf{P}^0(\varrho)$ the set of classes thus obtained. By (3.31), (3.32) and (3.33), for $(\underline{p},\pi)$ and $(\underline{p}',\pi')$ in the same class,

$$\langle\phi', \phi\rangle^L_{(\underline{p},\pi)} = \langle\phi', \phi\rangle^L_{(\underline{p}',\pi')} .$$

For a class $\{(\underline{p},\pi)\}$, denote by $\Phi_{\{(\underline{p},\pi)\}}$ a (finite) collection of measurable functions $\Phi_{(\underline{p},\pi)} : Z(\underline{p}) \to I\left(J_{(\underline{p},\pi)}\right)$, $(\underline{p},\pi) \in \{(\underline{p},\pi)\}$, such that (3.32) holds for all $\tau \in \mathfrak{S}_m$ and such that

$$\left\langle\left\langle\Phi_{\{(\underline{p},\pi)\}}, \Phi_{\{(\underline{p},\pi)\}}\right\rangle\right\rangle = \sum_{(\underline{p},\pi)\in\{(\underline{p},\pi)\}} c_{\underline{p}} \int_{Z(\underline{p})} \left\langle\left\langle\Phi_{(\underline{p},\pi)}(\underline{z}), \Phi_{(\underline{p},\pi)}(\underline{z})\right\rangle\right\rangle d_{\underline{p}}\underline{z} \tag{3.34}$$

is finite. Denote by $L_{\{(\underline{p},\pi)\}}$ the set of such collections.

For $\phi \in I(\rho)$, put $R_{\{(\underline{p},\pi)\}}\phi = \left\{R_{(\underline{p},\pi)}\phi : (\underline{p},\pi) \in \{(\underline{p},\pi)\}\right\}$. Similarly we define $R_{\{(\underline{p},\pi)\}}\phi'$ for $\phi' \in I(\rho')$. For $\{(\underline{p},\pi)\} \in \mathbf{P}^0(\varrho)$, the map $R_{\{(\underline{p},\pi)\}}$ has value in $L_{\{(\underline{p},\pi)\}}$.

We can remove L in (3.27) and (3.31) as in the general theory on Eisenstein series and get the following theorem.

THEOREM 3.15. (refer to [**MW89**, p. 649])
(1) Suppose ρ and ρ' are irreducible unitary representations of $\overline{T}$ such that $\rho, \rho' \in \varrho$. If $\phi \in I(\rho), \phi' \in I(\rho')$, we have

$$\langle\phi', \phi\rangle = \sum_{\{(\underline{p},\pi)\}\in\mathbf{P}^0(\varrho)} \left\langle\left\langle R_{\{(\underline{p},\pi)\}}\phi', R_{\{(\underline{p},\pi)\}}\phi\right\rangle\right\rangle .$$

(2) The space $\oplus_{\{(\underline{p},\pi)\}\in\mathbf{P}^0(\varrho)} L_{\{(\underline{p},\pi)\}}$ is the completion of the images of the maps $R_{\{(\underline{p},\pi)\}}I(\rho)$ when ρ runs through ϱ and $\{(\underline{p},\pi)\}$ passes $\mathbf{P}^0(\varrho)$.

PROOF. The theorem follows from the general theory of Eisenstein series in [**Lan76**, section 7]. See [**MW89**, p. 649] for details on the procedure for taking residues.

It is also convenient to prove the theorem by the results in [**MW93**] which states Langlands' theory in a more explicit way. By (3.27), the hypotheses in

[**MW93**, V.3.1.] are satisfied. It then follows from [**MW93**, V.3.2.(5)(ii)] that

$$\text{For } \{(\underline{p},\pi)\} \in \mathbf{P}^0(\varrho), \text{ the map } R_{\{(\underline{p},\pi)\}} \text{ has image in } L_{\{(\underline{p},\pi)\}} \tag{3.35}$$
$$\text{and } \langle\langle\cdot,\cdot\rangle\rangle \text{ is positive hermitian.}$$

Rewrite (3.27) as

$$\langle\phi',\phi\rangle =_L \sum_{\substack{\{(\underline{p},\pi)\}\\ \in\mathbf{P}^0(\varrho)}} \sum_{\substack{(\underline{p},\pi)\\ \in\{(\underline{p},\pi)\}}} \left(c_{\underline{p}} \int_{Z(\underline{p})}^{L} \left\langle\left\langle R_{(\underline{p},\pi)}\phi'(\underline{z}), R_{(\underline{p},\pi)}\phi(\underline{z})\right\rangle\right\rangle d_{\underline{p}}\underline{z} \right). \tag{3.36}$$

By [**MW93**, V.3.13.], we can remove L. Part (1) follows.

Part (2) is [**MW93**, V.3.14. corollary (ii)]. □

Denote by Θ_0 (refer to page 42) the cuspidal datum (T,ϱ) where ϱ is an "equivalence class up to conjugates and twists" of irreducible representations of $\overline{T}$.

Corollary 3.16. (refer to [**MW89**, p. 649])

$$L^2\left(\overline{G},\xi\right)_{\Theta_0} \cong \oplus_{\{(\underline{p},\pi)\}\in\mathbf{P}^0(\varrho)} L_{\{(\underline{p},\pi)\}}.$$

This answers the question at the end of section 3.3. We may visualize $L_{\{(\underline{p},\pi)\}}$ as follows. Fix a $(\underline{p},\pi) \in \{(\underline{p},\pi)\}$. The group $\mathfrak{S}_r$ acts on the space $Z_0(\underline{p})$ (defined by (3.30)) by permuting coordinates. Denote

$$\mathfrak{S}(\underline{p},\pi) = \left\{\sigma \in \mathfrak{S}_r : (\sigma\underline{p},\sigma\pi) = (\underline{p},\pi)\right\}.$$

Then

$$L_{\{(\underline{p},\pi)\}} = m(\overline{T}) \cdot \int_{Z_0(\underline{p})/\mathfrak{S}(\underline{p},\pi)}^{\oplus} i_{\overline{M}}^{\overline{G}} \left(J_{(\underline{p},\pi)}[\underline{t}]\right) d\underline{t}$$

where the integer $m(\overline{T})$ is the cardinality of $(T(F)\mathfrak{Z}(\overline{T})\backslash\overline{A'})^*$, and the meaning of $Z_0(\underline{p})/\mathfrak{S}(\underline{p},\pi)$ should be clear.

In particular, if $\sigma\varrho \neq \varrho$ for some $\sigma \in \mathfrak{S}_r$, then the discrete part is 0.

Otherwise, for some $\rho \in \varrho$, we have $\sigma\rho = \rho$, $\forall\sigma \in \mathfrak{S}_r$. Then the discrete part of $L^2\left(\overline{G},\xi\right)_{\{\Theta_0\}}$ is the sum of $m(\overline{T})$ copies of the unique irreducible quotient of $i_{\overline{T}}^{\overline{G}}\rho[\underline{\lambda}]$ where

$$\underline{\lambda} = ((N-1)/2n, (N-3)/2n, \cdots, (1-N)/2n).$$

In particular, by Corollary 3.10, if the covering satisfies condition (3.3), then the multiplicity $m(\overline{T})$ of the above quotient is 1. Remark that in this case,

$$\rho = L^2(\overline{T},\chi_T),$$

for the central character χ_T of ρ (see Corollary 3.11).

CHAPTER 4

CONTOUR INTEGRATION (after MW)

We prove lemma 3.13 in this chapter. Notice that we just use the notations and the method of [**MW89**]. However, to check that each step in [**MW89**, part III] works for the metaplectic group after replacing 1 by $1/n$ at suitable places, we record some sections of [**MW89**, part III] in detail.

Recall that for $i, j \leq r$, ρ_{ij}^n is a character of $\mathbf{A}^\times$ given by $\rho_{ij}^n(x) = \rho(h_{ij}(x^n))$ for any $x \in \mathbf{A}^\times$.

Fix ρ, ρ', ϕ, ϕ' and L as in the context of lemma 3.13. Recall that is ρ is an irreducible representation of $\overline{T}$, V_ρ is the isotropic subspace of ρ in $L^2(\overline{T}, \chi_T)$ where χ_T is the central character of ρ. Generally, we don't have multiplicity one for $L^2(\overline{T}, \chi_T)$, so V_ρ is not the representation space of ρ. Recall that $\phi \in V_\rho$. Note

$$\phi(\underline{s}) \in I(\rho, \underline{s}) \simeq i_{\overline{T}}^{\overline{G}} V_\rho[\underline{s}] \simeq m(\overline{T}) \cdot i_{\overline{T}}^{\overline{G}} \rho[\underline{s}].$$

Our local statement in Chapter 2, such as Proposition 2.8, do not apply directly. However, to show Lemma 3.13, it is enough to show (3.27) holds for $\phi(\underline{s}) \in i_{\overline{T}}^{\overline{G}} \rho[\underline{s}]$. The similar observation applies to ϕ'. So in this chapter, assume $\phi \in \mathbf{I}(\rho)$ and $\phi' \in \mathbf{I}(\rho')$ such that

$$\phi(\underline{s}) \in i_{\overline{T}}^{\overline{G}} \rho[\underline{s}], \qquad \phi'(\underline{s}) \in i_{\overline{T}}^{\overline{G}} \rho'[\underline{s}].$$

4.1. Holomorphy at the Origin of a Singular Hyper-plane

We keep the notation in section 3.5.

LEMMA 4.1. (refer to [**MW89**, p. 650])
Suppose $(\underline{p}, \pi) \in \mathbf{P}(\varrho)$, $\sigma \in \mathfrak{S}\left(\pi_{\underline{p}}, \rho\right)$, $\tau \in \mathfrak{S}\left(\pi_{\underline{p}}, \rho'\right)$:
1) The functions

$$N\left(w_{\underline{p}}, \underline{s}\right) M\left(\sigma^{-1}, \sigma\underline{s}\right) \phi\left(\sigma\underline{s}\right);$$

$$N\left(w_{\underline{p}}, -w_{\underline{p}}\underline{s}\right) M\left(\tau^{-1}, -\tau w_{\underline{p}}\underline{s}\right) \phi'\left(-\tau w_{\underline{p}}\underline{s}\right)$$

are holomorphic at a general point of $V\left(\underline{p}\right)$.
2) They are zero if $\sigma \notin \mathfrak{S}\left(\uparrow, \underline{p}\right)$, $\tau \notin \mathfrak{S}\left(\uparrow, \underline{p}\right)$.
3) Suppose $\sigma \in \mathfrak{S}\left(\uparrow, \underline{p}\right)$, $\tau \in \mathfrak{S}\left(\uparrow, \underline{p}\right)$. *Then*

$$M\left(\sigma^{-1}, \sigma\underline{s}\right) \phi\left(\sigma\underline{s}\right); \quad \textit{and} \quad M\left(\tau^{-1}, -\tau w_{\underline{p}}\underline{s}\right) \phi'\left(-\tau w_{\underline{p}}\underline{s}\right)$$

are holomorphic at a general point of $V\left(\underline{p}\right)$. □

This lemma permits us to define the above functions as meromorphic functions on $V\left(\underline{p}\right)$. The proof of this lemma is the same as [**MW89**] in the nonmetaplectic case.

Suppose $A_i = \{a_i + j/n : j \in \mathbf{Z}, 0 \leq j \leq l_i\}$ $(i = 1, 2)$ are two segments. We say that A_1 and A_2 are linked (in the global case) if neither contains the other and the union of them is again a segment.

Suppose $x \in V^0(\underline{p}) \cap \mathbf{R}^r$, $\underline{s} \in V(\underline{p})$. We are going to use the following notations:

$z^0 = (z_1^0, \cdots, z_m^0)$ the element in $\mathbf{R}^m$ such that $z_k^0 = x_i$ for $i \in \Delta_k$;
$E_k^x =$ the segment $\{z_k^0 + (1 - p_k)/2n, \cdots, z_k^0 + (p_k - 1)/2n\}$;
$\mathbf{E}^x$: the set of all pairs (k, h) such that
1) $1 \leq k < h \leq r$
2) $(ij)\pi_{\underline{p}} = \pi_{\underline{p}}, \forall i \in \Delta_k, j \in \Delta_h$
3) the segments E_k^x and E_h^x are linked;

$z(\underline{s}) = (z_1, \cdots, z_m)$ the element in $\mathbf{C}^m$ such that $z_k = s_i - \lambda(\underline{p})_i$ for $i \in \Delta_k$;
$V_x = \{\underline{s} \in V(\underline{p}) : \mathrm{Re}\underline{s} = \lambda(\underline{p}) + x, \|\mathrm{Im}(\underline{s})\| \leq 2T\}$;
$e^x(\underline{s}) = \prod_{(k,h)\in\mathbf{E}^x} (z_k - z_k^0 - z_h + z_h^0)$ a function on $V(\underline{p})$.
η: a fixed number such that $0 < \eta < 1/2n$ and such that for $i, j = 1, \cdots, r$ the function $L(s, \rho_{ij}^n)$ has no zero in the domain $\{s \in \mathbf{C} : 1 - n\eta < \mathrm{Re}(s) \leq 1, \|\mathrm{Im}(s)\| \leq 3T\}$.

For simplicity, let

$$\pi_{hk}^n = \left(\pi_{\underline{p}}\right)_{ij}^n,$$

for any $i \in \Delta_k$ and $j \in \Delta_h$. Note it is a unitary character of $\mathbf{A}^\times$.

LEMMA 4.2. (refer to [**MW89**, p. 651])
Suppose $(\underline{p}, \pi) \in \mathbf{P}(\underline{\varrho})$, $x \in V^0(\underline{p}) \cap \mathbf{R}^r$. Suppose the following conditions are satisfied:

if $1 \leq i < j \leq r$, $x_i - x_j > -\eta$.

Then the following function on $V(\underline{p})$

$$e^x(\underline{s}) \sum_{\tau\in\mathfrak{S}(\pi_{\underline{p}},\rho')\cap\mathfrak{S}(\uparrow,\underline{p})} N\left(w_{\underline{p}}, -w_{\underline{p}}\underline{s}\right) M\left(\tau^{-1}, -\tau w_{\underline{p}}\underline{s}\right) \phi\left(-\tau w_{\underline{p}}\underline{s}\right)$$

is holomorphic in a neighborhood of V_x in $V(\underline{p})$.

If furthermore $x_i - x_j \notin (1/2n)\mathbf{Z}$ when i, j belong to different intervals, then each term in the above sum is holomorphic in a neighborhood of V_x in $V(\underline{p})$.

PROOF. By considering the adjoint operator, we need to study

$$A(\underline{s}) = e^x(\underline{s}) \sum_{\tau\in\mathfrak{S}(\uparrow,\underline{p})\cap\mathfrak{S}(\pi_{\underline{p}},\rho')} M\left(\tau, w_{\underline{p}}\underline{s}\right) N\left(w_{\underline{p}}, \underline{s}\right) \psi(\underline{s})$$

for a function $\psi \in I\left(\pi_{\underline{p}}\right)$ which is K_v^*-invariant if $v \notin S$, where S is a finite set of places containing all archimedean ones. For $\tau \in \mathfrak{S}(\uparrow, \underline{p})$, the restriction of $N\left(\tau, w_{\underline{p}}\underline{s}\right)$ to the image of $N\left(w_{\underline{p}}, \underline{s}\right)$ is holomorphic in a neighborhood of V_x by Proposition 2.8. Calculate $r\left(\tau, w_{\underline{p}}\underline{s}\right)$. For $k, h = 1, \cdots, r$, put

$$r(k, h, \underline{s}) = \prod L\left(n(s_i - s_j), \pi_{kh}^n\right) L\left(n(s_i - s_j) + 1, \pi_{kh}^n\right)^{-1},$$

where the product is over $i \in \Delta_k, j \in \Delta_h$ such that $\left(w_{\underline{p}}i, w_{\underline{p}}j\right) \in \operatorname{inv}(\tau)$. Since $\tau \in \mathfrak{S}\left(\uparrow, \underline{p}\right)$, we have

$$r\left(\tau, w_{\underline{p}}\underline{s}\right) = \epsilon\left(\underline{s}\right) \prod_{1 \leq k < h \leq r} r\left(k, h, \underline{s}\right),$$

where $\epsilon\left(\underline{s}\right)$ collects all ϵ factors and possible a minus sign. Put $\tau' = \tau w_{\underline{p}}$. For $i \in \Delta_k,\ h > k$, put

$$\alpha\left(i, h\right) = \min\{j : j \in \Delta_h, \tau'\left(i\right) > \tau'\left(j\right)\}$$

if this set is not empty. Denote by $I_{k,h}$ the set of all i such that $\alpha\left(i,h\right) \neq \emptyset$ (we alway assume $k < h$). We see

$$\text{(4.1)} \quad r\left(k, h, \underline{s}\right) = \prod_{i \in I_{k,h}} L\left(n\left(s_i - s_{\alpha(i,h)}\right), \pi_{kh}^n\right) L\left(n\left(s_i - s_{p'_{h+1}}\right) + 1, \pi_{kh}^n\right)^{-1}.$$

For $j \in \Delta_h$, put

$$\beta\left(j, k\right) = \max\{i : i \in \Delta_k, \tau'\left(i\right) > \tau'\left(j\right)\}.$$

Denote by $J_{h,k}$ all such j that $\beta\left(j,k\right) \neq \emptyset$, then

$$\text{(4.2)} \quad r\left(k, h, \underline{s}\right) = \prod_{j \in J_{h,k}} L\left(n\left(s_{\beta(j,k)} - s_j\right), \pi_{kh}^n\right) L\left(n\left(s_{p'_k+1} - s_j\right) + 1, \pi_{kh}^n\right)^{-1}.$$

Suppose $p_h \geq p_k$. For $i \in I_{k,h}$ and $\underline{s}$ in a neighborhood of V_x, we have

$$\operatorname{Re}\left(s_i - s_{p'_{h+1}}\right) \geq \operatorname{Re}\left(s_{p'_{k+1}} - s_{p'_{h+1}}\right) > -\eta.$$

In the formula (4.1), the denominators are not zero. The poles are given by the numerators. There is a pole only if $(kh)\pi = \pi$, and if this is the case, the poles are on the hyper-plane $s_i - s_j = 1/n$ or 0. If $p_k > p_h$, we use (4.2) to get the same result.

Summing up, we can choose a neighborhood V of V_x such that the singular hyper-planes cut V only if $z_k^0 + p_k/2n \equiv z_h^0 + p_h/2n (\operatorname{mod} \frac{1}{n}\mathbf{Z})$. In particular, the last statement of the lemma follows. Denote by C the set of all pairs (k,h) verifying the above conditions and $k < h$, $(kh)\pi \cong \pi$.

In the general case, introduce $z\left(\underline{s}\right) = \left(z_1, \cdots, z_m\right) \in \mathbf{C}^m$. It parameterizes $V\left(\underline{p}\right)$. The above calculation shows that there is a function $n : C \to \mathbf{N}$ such that

$$\left(\prod_{(k,h) \in C} \left(z_k - z_k^0 - z_h + z_h^0\right)^{n(k,h)}\right) A\left(\underline{s}\right)$$

is holomorphic in a neighborhood of V_x.

Suppose $(k,h) \in C$, denote by $H_{k,h}$ the hyper-plane defined by the equation $\left(z_k - z_k^0 - z_h + z_h^0\right) = 0$. To finish the lemma, we only need to demonstrate

(4.3) There is a neighborhood V of V_x such that $A\left(\underline{s}\right)$ is holomorphic at a general point of $V \cap H_{k,h}$.

We choose V such that for $\underline{s} \in V$, we have

$$\text{if } 1 \leq i < j \leq r,\ \operatorname{Re}\left(s_i\right) - \lambda\left(\underline{p}\right)_i - \operatorname{Re}\left(s_j\right) + \lambda\left(\underline{p}\right)_j > -\eta,\ \|\operatorname{Im}\underline{s}\| < 3T.$$

By a general point of $V\cap H_{k,h}$ we mean a point such that for $(k',h')\neq(k,h)$, $i\in\Delta_{k'}$, $j\in\Delta_{h'}$, $L\left(n\left(s_i-s_j\right),\pi^n_{k'h'}\right)L\left(n\left(s_i-s_j\right)+1,\pi^n_{k'h'}\right)^{-1}$ does not have a pole. Put

$$\begin{aligned} J_0 &= \{(i,j): i\in\Delta_k, j\in\Delta_h, \lambda\left(\underline{p}\right)_i + x_i - \lambda\left(\underline{p}\right)_j - x_j = 0\} \\ J_1 &= \{(i,j): i\in\Delta_k, j\in\Delta_h, \lambda\left(\underline{p}\right)_i + x_i - \lambda\left(\underline{p}\right)_j - x_j = 1/n\}. \end{aligned}$$

Fix a general point $\underline{s}$ of $H_{k,h}\cap V$. Suppose $w'\in\mathfrak{S}_r$ such that if $w'^{-1}\underline{s}=\underline{s}'$ then $\mathrm{Re}\left(s'_i-s'_j\right)>-\eta$ for $1\le i<j\le r$ (this w' depends only on $z\left(\underline{s}\right)$). For $(i,j)\in J_0$, we have $s_i=s_j$. We can hence impose the supplementary condition $w'^{-1}j=w'^{-1}i+1$.

Put $\pi'_{\underline{p}}=w'^{-1}\pi_{\underline{p}}$. Applying Proposition 2.8, we can choose a neighborhood $V\left(\underline{s}\right)$ of $\underline{s}$ in $V\left(\underline{p}\right)$ and a holomorphic function $\psi': w'^{-1}V\left(\underline{s}\right)\to I\left(\pi'_{\underline{p}}\right)$, K^*-finite and invariant by K^*_v for $v\notin S$ such that

$$N\left(w_{\underline{p}},y\right)\psi\left(\underline{y}\right)=N\left(w_{\underline{p}}w',\underline{y}'\right)\psi'\left(\underline{y}'\right)\qquad\forall\underline{y}\in V\left(\underline{s}\right),\underline{y}'=w'^{-1}\underline{y}\,.$$

Choose a neighborhood $X\left(\underline{s}'\right)$ of $\underline{s}'$ in $X\left(T\right)\subset\mathbf{C}^r$ so that $X\left(\underline{s}'\right)\cap w'^{-1}V\left(\underline{p}\right)=w'^{-1}V\left(\underline{s}\right)$ and that ψ' extends to a holomorphic function $\psi': X\left(\underline{s}'\right)\to I\left(\pi'_{\underline{p}}\right)$ possessing the same invariant properties as the initial one. We can further suppose that $\mathrm{Re}\left(y'_i-y'_j\right)>-\eta$ for $1\le i<j\le N$, $\underline{y}'\in X\left(\underline{s}'\right)$.

If $\tau\in\mathfrak{S}\left(\uparrow,\underline{p}\right)$ and $\underline{y}\in V\left(\underline{s}\right)$, we have ($N_\tau$ is defined as in Proposition 2.8)

$$\begin{aligned} & M\left(\tau,w_{\underline{p}}\underline{y}\right)N\left(w_{\underline{p}},\underline{y}\right)\psi\left(\underline{y}\right) \\ &= r\left(\tau,w_{\underline{p}}\underline{y}\right)N_\tau\left(z\left(\underline{y}\right)\right)N\left(w_{\underline{p}},\underline{y}\right)\psi\left(\underline{y}\right) \\ &= r\left(\tau,w_{\underline{p}}\underline{y}\right)N_\tau\left(z\left(\underline{y}\right)\right)N\left(w_{\underline{p}}w',\underline{y}'\right)\psi'\left(\underline{y}'\right) \\ &= r\left(\tau,w_{\underline{p}}w'\underline{y}'\right)N\left(\tau w_{\underline{p}}w',\underline{y}'\right)\psi'\left(\underline{y}'\right). \end{aligned}$$

We need to study the analytic behavior of the following function in $w'^{-1}V\left(\underline{s}\right)$.

$$B\left(\underline{y}'\right)=e^x\left(\underline{y}\right)\sum_{\tau\in\mathfrak{S}\left(\pi_{\underline{p}},\rho'\right)\cap\mathfrak{S}\left(\uparrow,\underline{p}\right)}r\left(\tau,w_{\underline{p}}w'\underline{y}'\right)N\left(\tau w_{\underline{p}}w',\underline{y}'\right)\psi'\left(\underline{y}'\right).$$

Let $S\left(k,h\right)$ be the subgroup of $\mathfrak{S}_r$ generated by permutations exchanging two components of a pair $(i,j)\in J_0$. Put

$$W'=\left(\mathfrak{S}\left(\uparrow,\underline{p}\right)S\left(k,h\right)\right)\cap\mathfrak{S}\left(\pi_{\underline{p}},\rho'\right).$$

Define a meromorphic function on $X\left(\underline{s}'\right)$ by

$$C\left(\underline{y}'\right)=\sum_{\tau\in W'}r\left(\tau,w_{\underline{p}}w'\underline{y}'\right)N\left(\tau w_{\underline{p}}w',\underline{y}'\right)\psi'\left(\underline{y}'\right).$$

Now we prove that $B\left(\underline{y}'\right)$ is the restriction to $C\left(\underline{y}'\right)$ on $w'^{-1}V\left(\underline{s}\right)$ multiplied by $e^x\left(\underline{y}\right)$. It is enough to show that the terms corresponding to $\tau\notin\mathfrak{S}\left(\uparrow,\underline{p}\right)$ are zero on $w'^{-1}V\left(\underline{s}\right)$. The term $N\left(\tau w_{\underline{p}}w',\underline{y}'\right)\psi'\left(\underline{y}'\right)$ is holomorphic by lemma 2.7. But $r\left(\tau,w_{\underline{p}}w'\underline{y}'\right)$ is zero at a general point in $w'^{-1}V\left(\underline{s}\right)$ when $\tau\notin\mathfrak{S}\left(\uparrow,\underline{p}\right)$. The assertion follows.

We now study the behavior of $C\left(\underline{y}'\right)$. The term $N\left(\tau w_{\underline{p}}w', \underline{y}'\right)\psi'\left(\underline{y}'\right)$ is holomorphic. For $\tau \in \mathfrak{S}_r$,

$$r\left(\tau, w_{\underline{p}}w'\underline{y}'\right) = \epsilon\left(\underline{y}\right) \prod_{1\leq k'\leq h'\leq m} \prod \frac{L\left(n\left(y_i - y_j\right), \pi^n_{k'h'}\right)}{L\left(n\left(y_i - y_j\right) + 1, \pi^n_{k'h'}\right)}, \tag{4.4}$$

with the inner product over $i \in \Delta_{k'}, j \in \Delta_{h'}$, such that $\left(w_{\underline{p}}\left(i\right), w_{\underline{p}}\left(j\right)\right) \in \operatorname{inv}\left(\tau\right)$. All poles are simple and are given by $(i,j) \in J_1$. Hence

$$\left(\prod_{(i,j)\in J_1}\left(\left(y_i - y_j\right) - \frac{1}{n}\right)\right) r\left(\tau, w_{\underline{p}}w'\underline{y}'\right)$$

is holomorphic in $X\left(\underline{s}'\right)$ as well as the function

$$\left(\prod_{(i,j)\in J_1}\left(\left(y_i - y_j\right) - \frac{1}{n}\right)\right) C\left(\underline{y}'\right).$$

Suppose $(i,j) \in J_0$ and H is the hyper-plane of $\mathbf{C}^r$ defined by $\underline{y}' \in H \Leftrightarrow y_i = y_j$. We show that $C\left(\underline{y}'\right)$ is zero on $H \cap X\left(\underline{s}'\right)$.

It suffices to show this for a general point in $H \cap X\left(\underline{s}'\right)$. Denote by σ the elementary symmetry exchanging $w_{\underline{p}}\left(i\right)$ and $w_{\underline{p}}\left(j\right)$. Since $\left(w_{\underline{p}}\left(i\right), w_{\underline{p}}\left(j\right)\right) \in J_0$, $\sigma \in S\left(k,h\right)$. Put $\sigma' = \left(w_{\underline{p}}w'\right)^{-1}\sigma w_{\underline{p}}w'$. Then σ' is the elementary symmetry exchanging $w'^{-1}\left(i\right)$ and $w'^{-1}\left(i\right)+1$. Suppose $\tau \in W'$ and $\underline{y}'$ is a general point in $H \cap X\left(\underline{s}'\right)$. Note $\underline{y}'$ is fixed by σ' and $w_{\underline{p}}w'\underline{y}'$ is fixed by σ. We have

$$\begin{aligned} r\left(\tau\sigma, w_{\underline{p}}w'\underline{y}'\right) &= r\left(\tau, w_{\underline{p}}w'\underline{y}'\right) r\left(\sigma, w_{\underline{p}}w'\underline{y}'\right); \\ N\left(\tau\sigma w_{\underline{p}}w', \underline{y}'\right) &= N\left(\tau w_{\underline{p}}w', \underline{y}'\right) N\left(\sigma', \underline{y}'\right). \end{aligned}$$

For $r\left(\sigma, w_{\underline{p}}w'\underline{y}'\right)$, the pairs occurring in the product in (4.4) are

$$\left(w_{\underline{p}}\left(i\right), w_{\underline{p}}\left(j\right)\right), \quad \left(w_{\underline{p}}\left(i\right), e\right), \quad \left(e, w_{\underline{p}}\left(j\right)\right)$$

for $w_{\underline{p}}\left(i\right) < e < w_{\underline{p}}\left(j\right)$. Since the terms in (4.4) corresponding to the last two pairs are canceled, we have

$$r\left(\sigma, w_{\underline{p}}w'\underline{y}'\right) = r\left(\sigma', \underline{y}'\right).$$

By lemma 3.12, $M\left(\sigma', \underline{y}'\right) = -\mathrm{id}$. The above four identities imply that

$$r\left(\tau\sigma, w_{\underline{p}}w'\underline{y}'\right) N\left(\tau\sigma w_{\underline{p}}w', \underline{y}'\right)\psi'\left(\underline{y}'\right) = -r\left(\tau, w_{\underline{p}}w'\underline{y}'\right) N\left(\tau w_{\underline{p}}w', \underline{y}'\right)\psi'\left(\underline{y}'\right).$$

Hence $C\left(\underline{y}'\right) = -C\left(\underline{y}'\right)$ and we get what we wanted.

So there is a holomorphic function $D\left(\underline{y}'\right)$ on $X\left(\underline{s}'\right)$ such that

$$C\left(\underline{y}'\right) = \left(\prod_{(i,j)\in J_0}\left(y_i - y_j\right)\right)\left(\prod_{(i,j)\in J_1}\left(n\left(y_i - y_j\right) - 1\right)\right)^{-1} D\left(\underline{y}'\right).$$

The restriction of the above identity to $w'^{-1}V\left(\underline{s}'\right)$ is

$$B\left(\underline{y}'\right) = e^{x}\left(\underline{y}\right)\left(z_k - z_k^0 - z_h + z_h^0\right)^{|J_0|-|J_1|} D\left(\underline{y}'\right).$$

If E_k^x and E_h^x are not linked, then $|J_0| > |J_1|$. If E_h^x precedes E_k^x, then $|J_1| = |J_0| + 1$ and $(k,h) \in \mathbf{E}^x$ which implies that $\left(z_k - z_k^0 - z_h + z_h^0\right)$ divides $e^x\left(\underline{y}\right)$. So $B\left(\underline{y}'\right)$ is holomorphic in $w'^{-1}V\left(\underline{s}\right)$, whence (4.3). □

4.2. Corollaries

Suppose $\left(\underline{p},\pi\right) \in \mathbf{P}\left(\varrho\right)$ and $t \in \{0,\cdots,r\}$. We say that $\underline{p}$ is t-admissible if $p_i = 1$ for all $i = 1,\cdots,t$. We denote by $\mathbf{P}_t\left(\varrho\right)$ the set of all pairs $\left(\underline{p},\pi\right) \in \mathbf{P}\left(\varrho\right)$ such that $\underline{p}$ is T-admissible. Suppose $\left(\underline{p},\pi\right) \in \mathbf{P}_t\left(\varrho\right)$ and $x \in V\left(\underline{p}\right)^0 \cap \mathbf{R}^r$. We say that x is $\left(t,\underline{p}\right)$-admissible if $x_i - x_{i+1}$ is very large for $i \le t$ and $|x_i - x_j| < \eta$ for $t+1 \le i,j \le r$. Put

$$\mathfrak{S}_t\left(\uparrow,\underline{p}\right) = \{\sigma \in \mathfrak{S}\left(\uparrow,\underline{p}\right) : \sigma\left(i\right) = i \text{ for all } i = 1,\cdots,t\},$$

and for $k = t+1,\cdots,r$,

$$\mathfrak{S}_t^k\left(\uparrow,\underline{p}\right) = \{\sigma \in \mathfrak{S}_t\left(\uparrow,\underline{p}\right) : \sigma\left(p'_k+1\right) = t+1\}.$$

Lemma 4.3. (refer to [**MW89**, p. 656])

Suppose $t \in \{0,\cdots,r\}$ and $\left(\underline{p},\pi\right) \in \mathbf{P}_t\left(\varrho\right), x \in V^0\left(\underline{p}\right) \cap \mathbf{R}^r$ is a $\left(t,\underline{p}\right)$-admissible point. The following functions on $V\left(\underline{p}\right)$ are holomorphic in a neighborhood V_x in $V\left(\underline{p}\right)$:

$$\sum_{\sigma \in \mathfrak{S}_t^{t+1}\left(\uparrow,\underline{p}\right)\cap\mathfrak{S}\left(\pi_{\underline{p}},\rho\right)} N\left(w_{\underline{p}},\underline{s}\right) M\left(\sigma^{-1},\sigma\underline{s}\right)\phi\left(\sigma\underline{s}\right) \ ;$$

$$\sum_{\sigma \in \mathfrak{S}_t\left(\uparrow,\underline{p}\right)\cap\mathfrak{S}\left(\pi_{\underline{p}},\rho\right)} N\left(w_{\underline{p}},\underline{s}\right) M\left(\sigma^{-1},\sigma\underline{s}\right)\phi\left(\sigma\underline{s}\right) \ ;$$

$$\sum_{\tau \in \mathfrak{S}\left(\pi_{\underline{p}},\rho'\right)\cap\mathfrak{S}\left(\uparrow,\underline{p}\right)} N\left(w_{\underline{p}},-w_{\underline{p}}\underline{s}\right) M\left(\tau^{-1},-\tau w_{\underline{p}}\underline{s}\right)\phi\left(-\tau w_{\underline{p}}\underline{s}\right) .$$

If $x_i \ne x_j$ for any i,j belonging to different intervals, then each term in the above sums is holomorphic in V_x. □

Lemma 4.4. (refer to [**MW89**, p. 658])

Suppose $t \in \{0,\cdots,r\}$, $\left(\underline{p},\pi\right) \in \mathbf{P}_t\left(\varrho\right), x \in V^0\left(\underline{p}\right) \cap \mathbf{R}^r$. Suppose that

$x_i - x_{i+1}$ is very large for $i \le t$;
$|x_i - x_j| < \eta$ for $p'_{t+2} < i \le j \le r$;
$-\eta < x_i - x_j < 1/2n + \eta$ for $p'_{t+1} < i \le p'_{t+2} < j \le r$.

Then the following functions are holomorphic in a neighborhood of V_x in $V\left(\underline{p}\right)$:

$$\textstyle\sum_{\sigma \in \mathfrak{S}_t^{t+1}\left(\uparrow,\underline{p}\right)\cap\mathfrak{S}\left(\pi_{\underline{p}},\rho'\right)} N\left(w_{\underline{p}},\underline{s}\right) M\left(\sigma^{-1},\sigma\underline{s}\right)\phi\left(\sigma\underline{s}\right) \ ;$$

$$\textstyle\sum_{\tau \in \mathfrak{S}\left(\pi_{\underline{p}},\rho'\right)\cap\mathfrak{S}\left(\uparrow,\underline{p}\right)} N\left(w_{\underline{p}},-w_{\underline{p}}\underline{s}\right) M\left(\tau^{-1},-\tau w_{\underline{p}}\underline{s}\right)\phi\left(-\tau w_{\underline{p}}\underline{s}\right) .$$

Lemma 4.5. (refer to [**MW89**, p. 658])

Suppose $t \in \{0,\cdots,r\}$, $\left(\underline{p},\pi\right) \in \mathbf{P}_t\left(\varrho\right)$, $x \in V^0\left(\underline{p}\right) \cap \mathbf{R}^r$ is a $\left(t,\underline{p}\right)$-admissible point and $u \in \{t+1,\cdots,r\}$. The following function on $V\left(\underline{p}\right)$ is holomorphic in a neighborhood of V_x:

$$\sum_{\sigma,\tau}\left\langle N\left(w_{\underline{p}},-w_{\underline{p}}\underline{s}\right) M\left(\tau^{-1},-\tau w_{\underline{p}}\underline{s}\right)\phi'\left(-\tau w_{\underline{p}}\underline{s}\right), M\left(\sigma^{-1},\sigma\underline{s}\right)\phi\left(\sigma\underline{s}\right)\right\rangle$$

summing over $\sigma \in \mathfrak{S}_t^u(\uparrow, \underline{p}) \cap \mathfrak{S}\left(\pi_{\underline{p}}, \rho\right), \tau \in W(\uparrow, \underline{p}) \cap \mathfrak{S}\left(\pi_{\underline{p}}, \rho'\right)$. □

4.3. Residues by Induction

Suppose $t \in \{0, \cdots, r\}$, and $(\underline{p}, \pi) \in \mathbf{P}_t(\varrho), x \in V^0(\underline{p}) \cap \mathbf{R}^r$ is a $(t, \underline{p})$-admissible point. Put:

$$c_{\underline{p},t} = (2\pi)^{-d(\underline{p})} c'_{\underline{p}}/(r-t)!$$

Remark that $\mathfrak{S}_t(\uparrow, \underline{p}) = \cup_{u=t+1}^u \mathfrak{S}_t^u(\uparrow, \underline{p})$. Lemma 4.5 authorizes us to define

$$\langle \phi', \phi \rangle_{\underline{p},\pi,t}^L = c_{\underline{p},t} \int_{\underline{s} \in V(\underline{p}), \mathrm{Re}(\underline{s}) = \lambda(\underline{p}) + x}^L \sum_{\sigma,\tau}$$
$$\left\langle N\left(w_{\underline{p}}, -w_{\underline{p}}\overline{\underline{s}}\right) M\left(\tau^{-1}, -\tau w_{\underline{p}}\overline{\underline{s}}\right) \phi'\left(-\tau w_{\underline{p}}\overline{\underline{s}}\right), M\left(\sigma^{-1}, \sigma \underline{s}\right) \phi(\sigma \underline{s}) \right\rangle d_{\underline{p}}\underline{s},$$

summing over $\sigma \in \mathfrak{S}_t(\uparrow, \underline{p}) \cap \mathfrak{S}\left(\pi_{\underline{p}}, \rho\right)$, $\tau \in W(\uparrow, \underline{p}) \cap \mathfrak{S}\left(\pi_{\underline{p}}, \rho'\right)$.

PROPOSITION 4.6. (refer to [**MW89**, p. 659])
Suppose $t \in \{0, \cdots, m\}$, *we have*

$$\langle \phi', \phi \rangle =_L \sum_{(\underline{p},\pi) \in \mathbf{P}_t(\varrho)} \langle \phi', \phi \rangle_{\underline{p},\pi,t}^L .$$

When $t = 0$, *we get lemma 3.13.*

PROOF. The proof is by downward induction on t. If $t = m$, this is the definition of $\langle \cdot, \cdot \rangle$ (refer to (3.11)). Suppose the lemma is true for t; we show it for $t - 1$.

Fix $(\underline{p}, \pi) \in \mathbf{P}_t(\varrho)$ and x a $(t, \underline{p})$-admissible point such that $x_i \neq x_j$ if i, j belong to distinct intervals. We use this point to define $\langle \phi', \phi \rangle_{\underline{p},\pi,t}^L$. Introduce a point $y \in \mathbf{R}^r \cap X(T)$ such that $y_i - y_j = x_i - x_j$ if $i \neq t$ and $j \neq t$ but $|y_t - y_j| < \eta$ for $j \in \{t+1, \cdots, m\}$. Join $\lambda(\underline{p}) + y$ and $\lambda(\underline{p}) + x$ by a segment S. Let

$$I = \int_{\underline{s} \in V(\underline{p}), \mathrm{Re}(\underline{s}) = \lambda(\underline{p}) + y}^L \langle\langle B(\overline{\underline{s}}), A(\underline{s}) \rangle\rangle d_{\underline{p}}\underline{s},$$

where

$$A(\underline{s}) = \sum N\left(w_{\underline{p}}, \underline{s}\right) M\left(\sigma^{-1}, \sigma \underline{s}\right) \phi(\sigma \underline{s})$$
summing over $\sigma \in \mathfrak{S}_t(\uparrow, \underline{p}) \cap \mathfrak{S}\left(\pi_{\underline{p}}, \rho'\right)$;
$$B(\underline{s}) = \sum N\left(w_{\underline{p}}, -w_{\underline{p}}\underline{s}\right) M\left(\tau^{-1}, -\tau w_{\underline{p}}\underline{s}\right) \phi'\left(-\tau w_{\underline{p}}\underline{s}\right)$$
summing over $\tau \in \mathfrak{S}(\uparrow, \underline{p}) \cap \mathfrak{S}\left(\pi_{\underline{p}}, \rho'\right)$;

See (3.29) for the definition of $\langle\langle \cdot, \cdot \rangle\rangle$.

It is clear that $A(\underline{s})$ is holomorphic when $\mathrm{Re}(\underline{s}) \in S$. Keep the notation defined before Lemma 4.2 and apply Lemma 4.2 to $B(\underline{s})$. We see $\mathbf{E}^{x'} \neq \emptyset$ for x' such that $\lambda(\underline{p}) + x' \in S$ if and only if there is an h such that $t < h \leq r, \pi_t \cong \pi_h, x'_t =$

$x'_{p'_h+1} + (p_h + 1)/2n$. Such an h is unique. Denote by Ω the set of all h such that $t < h \le r, \pi_t \cong \pi_h$. The function

$$B(\underline{s}) \prod_{h \in P} \left(s_t - s_{p'_h+1} - \frac{1}{n} \right)$$

is holomorphic for $\mathrm{Re}(\underline{s}) \in S$. For $h \in \Omega$, define a function on $V(\underline{p})$:

$$C_h(\underline{s}) = \left(s_t - s_{p'_h+1} - \frac{1}{n} \right) \langle\langle B(\overline{\underline{s}}), A(\underline{s}) \rangle\rangle .$$

Suppose H_h is the hyper-plane defined by $s_t - s'_{p'_h+1} - 1/n = 0$. By the residue theorem

$$(4.5) \qquad \langle \phi', \phi \rangle^L_{\underline{p},\pi,t} - c_{\underline{p},t} I =_L 2\pi c_{\underline{p},t} \sum_{h \in \Omega} \int^L_{\substack{\underline{s} \in V(\underline{p}) \cap H_h \\ \mathrm{Re}(\underline{s}) = S \cap H_h}} C_h(\underline{s})\, d_h \underline{s} .$$

Fix $h \in \Omega$, suppose $v \in \mathfrak{S}_m$ such that

$$\begin{aligned} v: \quad & (1, \cdots, t, t+1, \cdots, h-1, h, h+1, \cdots, m) \\ \to \quad & (1, \cdots, t, t+2, \cdots, h, t+1, h+1, \cdots, m) . \end{aligned}$$

We identify v with an element in $\mathfrak{S}_r$ permuting intervals. Denote

$${}^h\underline{p} \in \mathbf{P} : {}^h\underline{p} = \left({}^h p_1, \cdots, {}^h p_{r-1} \right)$$

where ${}^h p_i = p_i$ for $i < h, i \ne t$, ${}^h p_i = p_{i+1}$ for $h \le i < r$, ${}^h p_t = p_h + 1$.

Denote by ${}^h\Delta_k$ the segment corresponding to ${}^h\underline{p}$. Remark that ${}^h\underline{p}$ is $(t-1)$-admissible. We have $v\left(V(\underline{p}) \cap H_h\right) = V\left({}^h\underline{p}\right)$.

Put ${}^h y = v(S \cap H_h) - \lambda\left({}^h\underline{p}\right)$; then

${}^h y_i - {}^h y_{i+1}$ is very large for $i \le t-1$;
$|{}^h y_i - {}^h y_j| = |x_{v^{-1}i} - x_{v^{-1}j}| < \eta$ if $p'_{t+1} \le i, j \le r$;
${}^h y_i - {}^h y_j = \frac{1}{2n} + x_{i'} - x_{v^{-1}j}$ if $i \in {}^h\Delta_t$, ${}^h p'_{t+1} \le j \le r$, where $i' \in \Delta_h$.

Denote by v' the element in $\mathfrak{S}_r$ such that it is the identity on ${}^h\Delta_k$ for $k \ne t$ and such that $v'(t+i) = t+i-1$ for $1 \le i \le {}^h p_t$; $v'(t) = {}^h p'_{t+1}$. We have

$$(4.6) \qquad v w_{\underline{p}} v^{-1} v'^{-1} = v' v w_{\underline{p}} v^{-1} = w_{{}^h\underline{p}} .$$

For the h-th integral in the sum (4.5), we change variables

$$\underline{s} = v^{-1}\underline{s}', \quad \sigma = \sigma' v, \quad \tau = \tau' v' v.$$

Then σ' runs over $\mathfrak{S}^t_{t-1}\left(\uparrow, {}^h\underline{p}\right) \cap \mathfrak{S}\left(\pi_{{}^h\underline{p}}, \rho\right)$. We may suppose τ is such that $\tau\left(p'_{h+1}\right) < \tau(t)$, otherwise the corresponding terms in $B(\underline{s})$ are holomorphic along H_h. This can be seen from the proof of lemma 4.2. Indeed, if $\tau\left(p'_{h+1}\right) > \tau(t)$, then in (4.1), $\alpha(i,h) \ne p'_h + 1$, hence there is no pole in (4.1) and we may ignore those terms. Then τ' goes over $\mathfrak{S}\left(\uparrow, {}^h\underline{p}\right) \cap \mathfrak{S}\left(\pi_{{}^h\underline{p}}, \rho\right)$. We have (observe $\tau w_{\underline{p}} = \tau' w_{{}^h\underline{p}} v$)

$$M\left(\sigma^{-1}, \sigma\underline{s}\right) \phi(\sigma\underline{s}) = M\left(v^{-1}, \underline{s}'\right) M\left(\sigma'^{-1}, \sigma'\underline{s}'\right) \phi\left(\sigma'\underline{s}'\right)$$

and

$$N\left(w_{\underline{p}},-w_{\underline{p}}\underline{s}\right)M\left(\tau^{-1},-\tau w_{\underline{p}}\underline{s}\right)\phi'\left(-\tau w_{\underline{p}}\underline{s}\right)$$
$$=\left(r\left(v^{-1},-v\underline{s}\right)r\left(v,-w_{\underline{p}}\underline{s}\right)r\left(v',-v'^{-1}w_{{}^h\underline{p}}\underline{s}'\right)\right)^{-1}$$
$$\times M\left(v^{-1},-\underline{s}'\right)N\left(w_{{}^h\underline{p}},-w_{{}^h\underline{p}}\underline{s}\right)M\left(\tau'^{-1},-\tau' w_{{}^h\underline{p}}\underline{s}'\right)\phi'\left(-\tau' w_{{}^h\underline{p}}\underline{s}'\right).$$

Since v permutes intervals, $r\left(v,-w_{\underline{p}}\underline{s}\right)=r\left(v,-\underline{s}\right)$. So the only r-factor left is

$$r\left(v'^{-1},-w_{{}^h\underline{p}}\underline{s}'\right). \tag{4.7}$$

Also (see (3.25) for the definition of $f_{\underline{p}}$)

$$f_{\underline{p}}\left(\underline{s}\right)\left(s_t-s_{p'_h+1}-\frac{1}{n}\right)r\left(v'^{-1},-w_{{}^h\underline{p}}\underline{s}'\right)r\left(w_{\underline{p}},\underline{s}\right)=f_{{}^h\underline{p}}\left(\underline{s}'\right)r\left(w_{{}^h\underline{p}},\underline{s}'\right).$$

Hence the value of the following function on $V\left(\underline{p}\right)\cap H_h$

$$\left(s_t-s_{p'_h+1}-\frac{1}{n}\right)\overline{r\left(v'^{-1},-w_{{}^h\underline{p}}\overline{\underline{s}}'\right)}$$

is $\overline{c'_{{}^h\underline{p}}c'_{\underline{p}}}^{-1}$.

By (4.7) and the adjoint formula (3.23), we obtain that the h-th integral of the formula (4.5) is

$$c'_{{}^h\underline{p}}c'^{-1}_{\underline{p}}\int_{\substack{\underline{s}\in V\left({}^h\underline{p}\right)\\ \mathrm{Re}(\underline{s})=\lambda\left({}^h\underline{p}\right)+{}^h\underline{y}}}\sum_{\sigma,\tau}$$
$$\left\langle N\left(w_{{}^h\underline{p}},-w_{{}^h\underline{p}}\overline{\underline{s}}\right)M\left(\tau^{-1},-\tau w_{{}^h\underline{p}}\overline{\underline{s}}\right)\phi'\left(-\tau w_{{}^h\underline{p}}\overline{\underline{s}}\right),M\left(\sigma^{-1},\sigma\underline{s}\right)\phi\left(\sigma\underline{s}\right)\right\rangle d_{{}^h\underline{p}}\underline{s}$$
$$\sigma\in\mathfrak{S}^t_{t-1}\left(\uparrow,{}^h\underline{p}\right)\cap\mathfrak{S}\left(\pi_{{}^h\underline{p}},\rho\right),\qquad \tau\in\mathfrak{S}\left(\uparrow,{}^h\underline{p}\right)\cap\mathfrak{S}\left(\pi_{{}^h\underline{p}},\rho'\right). \tag{4.8}$$

But ${}^h\underline{y}$ is not $\left(t-1,{}^h\underline{p}\right)$-admissible. So define ${}^hx\in V\left({}^h\underline{p}\right)\cap\mathbf{R}^r$ by

$$\begin{aligned} {}^hx_i-{}^hx_j={}^hy_i-{}^hy_j &\quad \text{if}\quad i\notin{}^h\Delta_t, j\notin{}^h\Delta_t;\\ \left|{}^hx_i-{}^hx_j\right|<\eta &\quad \text{if}\quad i\in{}^h\Delta_t, i<j. \end{aligned}$$

Denote the integral in (4.8) by I^h and by ${}^hI$ the same integral replacing ${}^hy$ by ${}^hx$. Suppose ${}^hS$ is the segment joining $\lambda\left({}^h\underline{p}\right)+{}^hx$ and $\lambda\left({}^h\underline{p}\right)+{}^hy$. By lemma 4.4 and lemma 4.5, $I^h=_L{}^hI$. So we have

$$\left\langle\phi',\phi\right\rangle^L_{\underline{p},\pi,t}=_L c_{\underline{p},t}I+\sum_{h\in P}2\pi c_{\underline{p},t}c'^{-1}_{\underline{p}}c'_{{}^h\underline{p}}\left({}^hI\right)$$

Suppose $u\in\{t,\cdots,r\}$ and $v\in\mathfrak{S}_r$ such that

$$\begin{aligned} v:\quad & (1,\cdots,t-1,t,t+1,\cdots,u,u+1,\cdots,r)\\ \mapsto\quad & (1,\cdots,t-1,u,t,\cdots,u-1,u+1,\cdots,r). \end{aligned}$$

Introduce the partition ${}_u\underline{p} = v\underline{p}$. As in lemma 4.5, I is equal to $I\left({}_u\underline{p}, u\right)$ where for a $(t-1)$-admissible partition $\underline{\tilde{p}} = (\tilde{p}_1, \cdots, \tilde{p}_{\tilde{m}})$ and $u \in \{t, \cdots, \tilde{m}\}$, we put

$$I\left(\underline{\tilde{p}}, u\right) = \int^L_{\substack{\underline{s}\in V\left(\underline{\tilde{p}}\right) \\ \mathrm{Re}(\underline{s})=\lambda\left(\underline{\tilde{p}}\right)+\tilde{x}}} \sum_{\sigma,\tau}$$

$$\left\langle N\left(w_{\underline{\tilde{p}}}, -w_{\underline{\tilde{p}}}\underline{\overline{s}}\right) M\left(\tau^{-1}, -\tau w_{\underline{\tilde{p}}}\underline{\overline{s}}\right) \phi'\left(-\tau w_{\underline{\tilde{p}}}\underline{\overline{s}}\right), M\left(\sigma^{-1}, \sigma\underline{s}\right) \phi(\sigma\underline{s})\right\rangle d_{\underline{\tilde{p}}}\underline{s}$$

summing over

$$\sigma \in \mathfrak{S}^u_{t-1}\left(\uparrow, \underline{\tilde{p}}\right) \cap \mathfrak{S}\left(\pi_{\underline{\tilde{p}}}, \rho\right), \quad \tau \in \mathfrak{S}\left(\uparrow, \underline{\tilde{p}}\right) \cap \mathfrak{S}\left(\pi_{\underline{\tilde{p}}}, \rho'\right),$$

$\tilde{x}$ being a $\left(t-1, \underline{\tilde{p}}\right)$-admissible point. One can write

$$I = (r-t+1)^{-1} \sum_{u=t}^{r} I\left({}_u\underline{p}, u\right).$$

Notice that $c_{\underline{p},t}(r-t+1)^{-1} = c_{{}_u\underline{p},t-1}$.

Similarly, for each ${}^hI$, if $v \in \mathfrak{S}_{r-1}$ such that

$$\begin{array}{rl} v: & (1, \cdots, t-1, t, t+1, \cdots, h-1, h, \cdots, r-1) \\ \longmapsto & (1, \cdots, t-1, h-1, t, \cdots, h-2, h, \cdots, r-1) \end{array}$$

and $\underline{p}^h = v^h\underline{p}$, we have ${}^hI = I\left(\underline{p}^h, h-1\right)$. Notice that $2\pi c_{\underline{p},t} c'^{-1}_{\underline{p}} c'_{{}^h\underline{p}} = c_{\underline{p}^h,t-1}$.

We obtain the equality

$$\langle\phi', \phi\rangle^L_{\underline{p},\pi,t} =_L \sum_{u=t}^{r} c_{{}_u\underline{p},t-1} I\left({}_u\underline{p}, u\right) + \sum_{h\in P} c_{\underline{p}^h,t-1} I\left(\underline{p}^h, h-1\right).$$

In the remainder of the proof, we use $r\left(\underline{p}\right), \Omega(\underline{p})$ instead of r, Ω to indicate the dependence on $\underline{p}$ (refer to page 58 for the definition of Ω) We have the bijection

$$\begin{array}{rl} & \left\{\left(\underline{p}, u\right) : \underline{p} \in \mathbf{P}_t(\varrho), t \le u \le r\left(\underline{p}\right)\right\} \cup \left\{\left(\underline{p}, h\right) : \underline{p} \in \mathbf{P}_t(\varrho), h \in \Omega(\underline{p})\right\} \\ \longleftrightarrow & \left\{\left(\underline{\tilde{p}}, k\right) : \underline{\tilde{p}} \in \mathbf{P}_{t-1}(\varrho), t \le k \le r\left(\underline{\tilde{p}}\right)\right\} \end{array}$$

given by $\left(\underline{p}, u\right) \mapsto \left({}_u\underline{p}, u\right)$ on the first set and $\left(\underline{p}, h\right) \mapsto \left(\underline{p}^h, h-1\right)$ on the second. Hence

$$\sum_{\underline{p}\in\mathbf{P}_t} \langle\phi', \phi\rangle^L_{\underline{p},\pi,t} =_L \sum_{\underline{\tilde{p}}\in\mathbf{P}_{t-1}} c_{\underline{\tilde{p}},t-1} \sum_{k=t}^{r\left(\underline{\tilde{p}}\right)} I\left(\underline{\tilde{p}}, k\right).$$

Since

$$\mathfrak{S}_{t-1}\left(\uparrow, \underline{\tilde{p}}\right) = \cup_{k=t}^{r\left(\underline{\tilde{p}}\right)} \mathfrak{S}^k_{t-1}\left(\uparrow, \underline{\tilde{p}}\right),$$

the expression after the first sum symbol on the right hand side is $=_L \langle\phi', \phi\rangle^L_{\underline{\tilde{p}},t-1}$.
Now the lemma follows. □

Bibliography

[Art79] J Arthur. Eisenstein series and the trace formula. In *Proceedings of Symposia in Pure Mathematics*, volume 33, pages 253–274, 1979. Part 1.

[BJ79] A Borel and H Jacquet. Automorphic forms and automorphic representations. In *Proceedings of Symposia in Pure Mathematics*, volume 33, pages 189–202, 1979. Part 1.

[BZ77] I N Bernstein and A V Zelevinsky. Induced representations of reductive p-adic groups. I. *Ann. scient. Éc. Norm. Sup.*, 10:441–472, 1977.

[Fla79] D Flath. Decomposition of representations into tensor products. In *Proceedings of Symposia in Pure Mathematics*, volume 33, pages 179–183, 1979. Part 1.

[Fli80] Y Z Flicker. Automorphic forms on covering groups of $GL(2)$. *Invent. Math.*, 57:119–182, 1980.

[KP84] D A Kazhdan and S J Patterson. Metaplectic forms. *Publ. Math. I.H.E.S.*, 59:35–142, 1984.

[KS88] C D Keys and F Shahidi. Artin l-functions and normalization of intertwining operators. *Ann. Scient. Ec. Norm. Sup.*, 4(21):67–89, 1988.

[Lan76] R P Langlands. *On the Functional Equations Satisfied by Eisenstein Series.* Springer Verlag, 1976.

[Mat69] H Matsumoto. Sur les sous-groupes arithmétiques des groupes semi-simples déployés. *Ann. Scient. Ec. Norm. Sup.*, 4(2):1–62, 1969.

[Moo68] C Moore. Group extensions of p-adic and adelic linear groups. *Publ. Math. I.H.E.S.*, 35:157–222, 1968.

[MW89] C Mœglin and J L Waldspurger. Le spectre résiduel de $GL(n)$. *Ann. Scient. Ec. Norm. Sup.*, 22(4):605–674, 1989.

[MW93] C Mœglin and J L Waldspurger. *Décomposition Spectrale et Séries d'Eisenstein (Une Paraphrase de l'Écriture).* Birkhäuse Verlag, 1993.

[Nag49] H Nagao. The extensions of topological groups. *Osaka Journal of Math*, 1:36–42, 1949.

[Ste62] R Steinberg. Générateurs, relations et revêtements de groupes algébriques. In *Colloque de Bruxelles*, pages 113–127, 1962.

[Sun] H Sun. Remarks on certain metaplectic groups. to appear in Canadian Mathematical Bulletin.

[Wei74] A Weil. *Basic Number Theory.* Springer Verlag, 1974.

Index

Editorial Information

To be published in the *Memoirs*, a paper must be correct, new, nontrivial, and significant. Further, it must be well written and of interest to a substantial number of mathematicians. Piecemeal results, such as an inconclusive step toward an unproved major theorem or a minor variation on a known result, are in general not acceptable for publication. Papers appearing in *Memoirs* are generally longer than those appearing in *Transactions*, which shares the same editorial committee.

As of November 30, 2001, the backlog for this journal was approximately 6 volumes. This estimate is the result of dividing the number of manuscripts for this journal in the Providence office that have not yet gone to the printer on the above date by the average number of monographs per volume over the previous twelve months, reduced by the number of volumes published in four months (the time necessary for preparing a volume for the printer). (There are 6 volumes per year, each containing at least 4 numbers.)

A Consent to Publish and Copyright Agreement is required before a paper will be published in the *Memoirs*. After a paper is accepted for publication, the Providence office will send a Consent to Publish and Copyright Agreement to all authors of the paper. By submitting a paper to the *Memoirs*, authors certify that the results have not been submitted to nor are they under consideration for publication by another journal, conference proceedings, or similar publication.

Information for Authors

Memoirs are printed from camera copy fully prepared by the author. This means that the finished book will look exactly like the copy submitted.

The paper must contain a *descriptive title* and an *abstract* that summarizes the article in language suitable for workers in the general field (algebra, analysis, etc.). The *descriptive title* should be short, but informative; useless or vague phrases such as "some remarks about" or "concerning" should be avoided. The *abstract* should be at least one complete sentence, and at most 300 words. Included with the footnotes to the paper should be the 2000 *Mathematics Subject Classification* representing the primary and secondary subjects of the article. The classifications are accessible from `www.ams.org/msc/`. The list of classifications is also available in print starting with the 1999 annual index of *Mathematical Reviews*. The Mathematics Subject Classification footnote may be followed by a list of *key words and phrases* describing the subject matter of the article and taken from it. Journal abbreviations used in bibliographies are listed in the latest *Mathematical Reviews* annual index. The series abbreviations are also accessible from `www.ams.org/publications/`. To help in preparing and verifying references, the AMS offers MR Lookup, a Reference Tool for Linking, at `www.ams.org/mrlookup/`. When the manuscript is submitted, authors should supply the editor with electronic addresses if available. These will be printed after the postal address at the end of the article.

Electronically prepared manuscripts. The AMS encourages electronically prepared manuscripts, with a strong preference for AMS-LaTeX. To this end, the Society has prepared AMS-LaTeX author packages for each AMS publication. Author packages include instructions for preparing electronic manuscripts, the *AMS Author Handbook*, samples, and a style file that generates the particular design specifications of that publication series. Though AMS-LaTeX is the highly preferred format of TeX, author packages are also available in AMS-TeX.

Authors may retrieve an author package from e-MATH starting from `www.ams.org/tex/` or via FTP to `ftp.ams.org` (login as `anonymous`, enter username as password, and type `cd pub/author-info`). The *AMS Author Handbook* and the *Instruction Manual* are available in PDF format following the author packages link from `www.ams.org/tex/`. The author package can be obtained free of charge by sending email to `pub@ams.org` (Internet) or from the Publication Division, American Mathematical Society, P.O. Box 6248, Providence, RI 02940-6248. When requesting an author package, please specify AMS-LaTeX or AMS-TeX, Macintosh or IBM (3.5) format, and the publication in which your paper will appear. Please be sure to include your complete mailing address.

Sending electronic files. After acceptance, the source file(s) should be sent to the Providence office (this includes any TeX source file, any graphics files, and the DVI or PostScript file).

Before sending the source file, be sure you have proofread your paper carefully. The files you send must be the EXACT files used to generate the proof copy that was accepted for publication. For all publications, authors are required to send a printed copy of their paper, which exactly matches the copy approved for publication, along with any graphics that will appear in the paper.

TeX files may be submitted by email, FTP, or on diskette. The DVI file(s) and PostScript files should be submitted only by FTP or on diskette unless they are encoded properly to submit through email. (DVI files are binary and PostScript files tend to be very large.)

Electronically prepared manuscripts can be sent via email to `pub-submit@ams.org` (Internet). The subject line of the message should include the publication code to identify it as a Memoir. TeX source files, DVI files, and PostScript files can be transferred over the Internet by FTP to the Internet node `e-math.ams.org` (130.44.1.100).

Electronic graphics. Comprehensive instructions on preparing graphics are available at `www.ams.org/jourhtml/graphics.html`. A few of the major requirements are given here.

Submit files for graphics as EPS (Encapsulated PostScript) files. This includes graphics originated via a graphics application as well as scanned photographs or other computer-generated images. If this is not possible, TIFF files are acceptable as long as they can be opened in Adobe Photoshop or Illustrator. No matter what method was used to produce the graphic, it is necessary to provide a paper copy to the AMS.

Authors using graphics packages for the creation of electronic art should also avoid the use of any lines thinner than 0.5 points in width. Many graphics packages allow the user to specify a "hairline" for a very thin line. Hairlines often look acceptable when proofed on a typical laser printer. However, when produced on a high-resolution laser imagesetter, hairlines become nearly invisible and will be lost entirely in the final printing process.

Screens should be set to values between 15% and 85%. Screens which fall outside of this range are too light or too dark to print correctly. Variations of screens within a graphic should be no less than 10%.

Inquiries. Any inquiries concerning a paper that has been accepted for publication should be sent directly to the Electronic Prepress Department, American Mathematical Society, P. O. Box 6248, Providence, RI 02940-6248.

Editors

Selected Titles in This Series

(Continued from the front of this publication)

713 **Dorina Mitrea, Marius Mitrea, and Michael Taylor,** Layer potentials, the Hodge Laplacian, and global boundary problems in nonsmooth Riemannian manifolds, 2001

712 **Raúl E. Curto and Woo Young Lee,** Joint hyponormality of Toeplitz pairs, 2001

711 **V. G. Kac, C. Martinez, and E. Zelmanov,** Graded simple Jordan superalgebras of growth one, 2001

710 **Brian Marcus and Selim Tuncel,** Resolving Markov chains onto Bernoulli shifts via positive polynomials, 2001

709 **B. V. Rajarama Bhat,** Cocylces of CCR flows, 2001

708 **William M. Kantor and Ákos Seress,** Black box classical groups, 2001

707 **Henning Krause,** The spectrum of a module category, 2001

706 **Jonathan Brundan, Richard Dipper, and Alexander Kleshchev,** Quantum Linear groups and representations of $GL_n(\mathbb{F}_q)$, 2001

705 **I. Moerdijk and J. J. C. Vermeulen,** Proper maps of toposes, 2000

704 **Jeff Hooper, Victor Snaith, and Min van Tran,** The second Chinburg conjecture for quaternion fields, 2000

703 **Erik Guentner, Nigel Higson, and Jody Trout,** Equivariant E-theory for C^*-algebras, 2000

702 **Ilijas Farah,** Analytic guotients: Theory of liftings for quotients over analytic ideals on the integers, 2000

701 **Paul Selick and Jie Wu,** On natural coalgebra decompositions of tensor algebras and loop suspensions, 2000

700 **Vicente Cortés,** A new construction of homogeneous quaternionic manifolds and related geometric structures, 2000

699 **Alexander Fel′shtyn,** Dynamical zeta functions, Nielsen theory and Reidemeister torsion, 2000

698 **Andrew R. Kustin,** Complexes associated to two vectors and a rectangular matrix, 2000

697 **Deguang Han and David R. Larson,** Frames, bases and group representations, 2000

696 **Donald J. Estep, Mats G. Larson, and Roy D. Williams,** Estimating the error of numerical solutions of systems of reaction-diffusion equations, 2000

695 **Vitaly Bergelson and Randall McCutcheon,** An ergodic IP polynomial Szemerédi theorem, 2000

694 **Alberto Bressan, Graziano Crasta, and Benedetto Piccoli,** Well-posedness of the Cauchy problem for $n \times n$ systems of conservation laws, 2000

693 **Doug Pickrell,** Invariant measures for unitary groups associated to Kac-Moody Lie algebras, 2000

692 **Mara D. Neusel,** Inverse invariant theory and Steenrod operations, 2000

691 **Bruce Hughes and Stratos Prassidis,** Control and relaxation over the circle, 2000

690 **Robert Rumely, Chi Fong Lau, and Robert Varley,** Existence of the sectional capacity, 2000

For a complete list of titles in this series, visit the AMS Bookstore at **www.ams.org/bookstore/**.